NANOPESTICIDES

A Perspective from Welfare to Risk Assessment

NANOPESTICIDES

A Perspective from Welfare to Risk Assessment

Garima Awasthi, PhD
Payal Lodha, PhD
Gajendra Pal Singh, PhD
Editors

AAP | APPLE ACADEMIC PRESS

First edition published 2026

Apple Academic Press Inc.
1265 Goldenrod Circle, NE,
Palm Bay, FL 32905 USA

760 Laurentian Drive, Unit 19,
Burlington, ON L7N 0A4, Canada

CRC Press
2385 NW Executive Center Drive,
Suite 320, Boca Raton FL 33431

4 Park Square, Milton Park,
Abingdon, Oxon, OX14 4RN, UK

Apple Academic Press exclusively co-publishes with CRC Press, an imprint of Taylor & Francis Group, LLC

Library and Archives Canada Cataloguing in Publication

CIP data on file with Canada Library and Archives

Library of Congress Cataloging-in-Publication Data

CIP data on file with US Library of Congress

ISBN: 978-1-77491-966-8 (hbk)
ISBN: 978-1-00360-587-4 (ebk)
DOI: 10.1201/9781003605874

About the Editors

Garima Awasthi, PhD
Professor, K. R. Mangalam University, Gurugram, Haryana, India

Garima Awasthi, PhD, is a Professor at K. R. Mangalam University, Gurugram, Haryana, India. She worked as an Associate Professor, Vivekananda Global University, Jaipur, Rajasthan, India. She earned her PhD in Biological Sciences from the CSIR-National Botanical Research Institute (CSIR-NBRI). She also worked as a SERB-National Postdoctoral Fellow at the University of Lucknow. In recent years, she has published more than 60 scientific articles in internationally indexed and peer-reviewed journals on heavy metal stress in plants, mainly focusing on arsenic problems in rice, which have been widely cited. She has been a reviewer of scientific articles and manuscripts for various international scientific publishers. The line of research that she develops is multidisciplinary, covering the areas of plant abiotic and biotic stress, nanotechnology, plant physiology, and proteomics. She has collaborated with various educational institutions and research centers, as well as with the industrial sector. Dr. Awasthi has taught various courses and seminars and has participated in numerous national and international scientific conferences. In addition, she has advised various students on the development and completion of bachelor's degrees, Master of Science degrees, and Doctor of Science theses in the life sciences discipline.

Payal Lodha, PhD
Full-Time Professor, Department of Botany, University of Rajasthan Jaipur, Rajasthan, India

Payal Lodha, PhD, is a full-time Professor in the Department of Botany at the University of Rajasthan, Jaipur, Rajasthan, India. She obtained her PhD in Botany from the University of Rajasthan, Jaipur, in 1991. She served as Pool Officer (Scientist, CSIR), Assistant Professor, and Associate Professor in the department before taking charge as Professor. She has more than 30 years of research experience in the field of plant-nematode interactions. She has published more than 60 research and review articles in high-impact international and national journals. Professor Lodha was also deputed by the University of Rajasthan in September 2000 to assist the Government of Rajasthan in organizing the International Rajasthani Conclave-2000 (IRC-2000). She was selected to present a paper at the 2004 World Congress on In Vitro Biology in San Francisco, California, USA, in May 2004. She is also a conferred Fellow of the Botanical Society, Treasurer of the Indian Science Congress Association (ISCA). Jaipur Chapter (2013–2015), and Convenor of the ISCA Jaipur Chapter (2015–2021). She is also serving as Convenor of the Board of Studies (Botany) at the University of Rajasthan, Jaipur. She has organized more than 10 national and international conferences. She is also a member of the Indian Botanical Society, the Indian Society of Mycology and Plant Pathology, the Indian Phytological Society, and the Indian Science Congress.

Gajendra Pal Singh, PhD
Professor, Department of Botany, University of Rajasthan Jaipur, Rajasthan, India

Gajendra Pal Singh, PhD, is currently working as a Professor in the Department of Botany at the University of Rajasthan, Jaipur, Rajasthan, India. Previously, he worked as an Associate Professor (2009 to 2012) and an Assistant Professor (1996 to 2009) in the Department of Botany at the University of Rajasthan, Jaipur. His area of research interest is the rapid response time of the algal community to changing environmental conditions and the effect of various mutagens on the growth and productivity of algal products. He has more than 30 years of teaching and research experience. He has published more than 100 research papers in peer-reviewed national and international journals of high repute. He has organized four international and three national conferences. Other than research, he is an active administrator and has held reputed positions such as Dean of Student Welfare, Convener of the Revaluation Committee, and Additional Coordinator of UCCS and IT at the University of Rajasthan, Jaipur, as well as Vice-Principal, Rector, and Proctor at the University Maharaja College, Jaipur. He is a member of international and national professional bodies such as the International Society for Life Sciences (ISLS)–executive member; Indian Botanical Society (IBS)–Fellow of the Indian Botanical Society; and the Indian Science Congress Association (ISCA)–life member. He has delivered 15 keynote lectures and invited talks at international and national conferences. More than 36 students have completed their master's dissertations and doctoral work under his exemplary guidance. He is an active board member and reviewer for national and international journals.

Contents

Contributors

Tushar R. Ahire
Chemical Sciences Division, ICMR–National Institute of Occupational Health, Ahmedabad, Gujarat, India; Department of Biochemistry and Forensic Science, Gujarat University, Ahmedabad, Gujarat, India

Vinay Aseri
Department of Forensic Science, Vivekananda Global University, Jaipur, Rajasthan, India

Anjali Awasthi
Department of Zoology, University of Rajasthan, Jaipur, Rajasthan, India

Garima Awasthi
Professor, K. R. Mangalam University, Gurugram, Haryana, India

Kumud Kant Awasthi
Professor, K. R. Mangalam University, Gurugram, Haryana, India

K. Chandrasekhar
Department of Biotechnology, Vignan's Foundation for Science, Technology, and Research, Vadlamudi, Guntur, Andhra Pradesh, India

Rushikesh L. Chopade
Department of Forensic Science, Vivekananda Global University, Jaipur, Rajasthan, India

Shalini Fagna
Department of Botany, University of Rajasthan, Jaipur, Rajasthan, India

Pavan Kumar Ganechary
Laboratory of Biology and Biotechnology, National Forensic Sciences University, Gandhinagar, Gujarat, India

Shivangi Giri
Department of Zoology, IIS Deemed to be University, Jaipur, Rajasthan, India

Vanisha Godara
Department of Forensic Science, Vivekananda Global University, Jaipur, Rajasthan, India

Tanisha Gupta
Laboratory of Biology and Biotechnology, National Forensic Sciences University, Gandhinagar, Gujarat, India

Divya Kottadiyil
School of Behavioural and Forensic Sciences, JSPM University, Pune, India

Poonam Kumari
Department of Forensic Science, Vivekananda Global University, Jaipur, Rajasthan, India

Supriya Kumari
Department of Forensic Science, Vivekananda Global University, Jaipur, Rajasthan, India

Surbhi Kumari
Department of Forensic Science, Vivekananda Global University, Jaipur, Rajasthan, India

Tarun Kumar Kumawat
Department of Botany, University of Rajasthan, Jaipur, Rajasthan, India

Payal Lodha
Department of Botany, University of Rajasthan, Jaipur, Rajasthan, India

Tara Maheshwari
Department of Botany, University of Rajasthan, Jaipur, Rajasthan, India

Archana Meena
Department of Botany, University of Rajasthan, Jaipur, Rajasthan, India

Poonam Meena
Department of Botany, University of Rajasthan, Jaipur, Rajasthan, India

Varad Nagar
Department of Forensic Science, Vivekananda Global University, Jaipur, Rajasthan, India

Harsh Pandey
Department of Biotechnology and Chemical Engineering, School of Civil and Chemical Engineering, Manipal University, Jaipur, Rajasthan, India

Kamakshi Pandey
Materials Research Center, Malaviya National Institute of Technology, Jaipur, Rajasthan, India; Department of Physics, Malaviya National Institute of Technology, Jaipur, Rajasthan, India

Apexa Pareek
Department of Botany, University of Rajasthan, Jaipur, Rajasthan, India

Vishwa Patel
Department of Botany, M.S. University of Baroda, Vadodara, Gujarat, India

Neelam Poonar
Department of Botany, University of Rajasthan, Jaipur, Rajasthan, India

Sukanya Sachdeva
Chemical Sciences Division, ICMR–National Institute of Occupational Health, Ahmedabad, Gujarat, India; Department of Biochemistry and Forensic Science, Gujarat University, Ahmedabad, Gujarat, India

Dev Saini
Department of Life Sciences, Vivekananda Global University, Jaipur, Rajasthan, India

Mahipal Singh Sankhla
Assistant Professor, K. R. Mangalam University, Gurugram, Haryana, India

Lata Shahani
Department of Zoology, IIS Deemed to be University, Jaipur, Rajasthan, India

Anuj Sharma
Department of Forensic Science, Vivekananda Global University, Jaipur, Rajasthan, India

Renu Sharma
Department of Botany, University of Rajasthan, Jaipur, Rajasthan, India

Sujata Sharma
Department of Life Sciences, Vivekananda Global University, Jaipur, Rajasthan, India

Vaibhav Sharma
Department of Biotechnology, Himachal Pradesh University, Shimla, Himachal Pradesh, India

Apoorva Singh
Department of Forensic Science, Vivekananda Global University, Jaipur, Rajasthan, India

G. P. Singh
Department of Botany, University of Rajasthan, Jaipur, Rajasthan, India

Meenakshi Singh
Department of Ecology and Biodiversity, Sustaina Greens LLP, Vadodara, Gujarat, India

Perumal Sivaperumal
Chemical Sciences Division, ICMR–National Institute of Occupational Health, Ahmedabad, Gujarat, India

Darshini Trivedi
Department of Botany, M.S. University of Baroda, Vadodara, Gujarat, India

Rajbala Verma
Department of Zoology, University of Rajasthan, Jaipur, Rajasthan, India

Chandra Shekhar Yadav
Forensic Biology and DNA Forensic Lab. National Forensic sciences University, Guwahati Campus, Assam, India

Abbreviations

AgNPs	silver nanoparticles
AI	active ingredient
APX	ascorbate peroxidase
BEN	Brunauer-Emmett-Teller
BFN	biological nitrogen fixation
BIS	Bureau of Indian Standards
BP	bulk particles
CDA	cellulose diacetate
CIB&RC	Central Insecticide Board and Registration Committee
CNPE	$Cu(OH)_2$ nanopesticide
CNPs	carbon nanoparticles
CNTs	carbon nanotubes
$Cu(OH)_2$	copper hydroxide nanopesticide
CAT	catalase
DSC	differential scanning calorimetry
DLS	dynamic light scattering
DHAR	dehydroascorbate reductase
ENPs	engineered nanoparticles
EDS	energy-dispersive x-ray spectroscopy
EU	European Union
FDA	Food and Drug Administration
FCO	The Fertilizer (Control) Order
FIFRA	Federal Insecticide, Fungicide, and Rodenticide Act
FSSAI	Food Safety and Standards Authority of India
FTIR	Fourier Transform Infrared Spectroscopy
GHSP	Global Horizon Scanning Project
GR	glutathione reductase
GST	glutathione-s-transferase
HPF	human pulmonary fibroblasts
IPM	integrated pest management
JA	jasmonic acid
LDH	layered double hydroxides
LH	layered hydroxides
LHS	layered hydroxide salts
LOEC	lowest-observed-effect concentration
MBC	minimum biocidal concentration

MNPs	metallic nanoparticles
NAPs	nano-agri products
NBR	nano bioremediation
NNI	national nanotechnology initiative
NP	nanopesticides
NPF	nanoparticle fertilizers
NPs	nanoparticles
OM	organic matter
PCBs	polychlorinated biphenyls
PDP	pesticide data program
PEG	polyethylene glycol
PGPR	plant growth-promoting rhizobacteria
PDI	polydispersity index
PLA	polylactic acid
PNE	permethrin nanoemulsion
PVA	polyvinyl alcohol
PVP	polyvinylpyrrolidone
ROS	reactive oxygen species
SA	salicylic acid
SEM	scanning electron microscopy
SOD	superoxide dismutase
TCE	trichloroethylene
TEM	transmission electron microscopy
WHO	world health organization
XRD	X-ray diffraction

Preface

Pesticides have been widely used in agriculture to protect crops from pests and increase productivity for decades. However, conventional pesticides can have adverse effects on human health and the environment. The development of nanotechnology has led to the emergence of a new generation of pesticides, known as nanopesticides, which are designed to be more effective and safer than conventional pesticides.

The very first chapter of the book describes nanopesticides in a nutshell. The next chapter deals with the fate of nanopesticides, particularly in the context of soil and water. The third chapter is about nanopesticide communication with soil microorganisms, which play a crucial role in nutrient cycling and maintaining soil health. The fourth, fifth, and sixth chapters discuss the configuration and preparation methods, categories of nanopesticides used for crop protection, and nanocarrier-based nanopesticide formulations. However, there are also concerns about their potential risks to human health and the environment, and the need for effective risk assessment and regulation is becoming increasingly urgent. The seventh chapter is about the interaction between plants and nanopesticides. They interact with plants and control pests through various mechanisms, including direct contact with plant tissues, ingestion by pests, and systemic uptake by plants. The eighth chapter is all about the economic and market status of nanopesticides. The ninth chapter focuses on the environmental risk assessment of nanopesticides. Last but not least, the tenth chapter describes the impact of nanopesticides on non-target organisms. The book shall be a wonderful resource for students, academicians, and the general public to gain knowledge about nanopesticides.

We would like to express our sincere thanks to all the authors who contributed to this book, as well as to the publishers for their support in bringing this important topic to the forefront. We hope this book will stimulate further discussion and research on the potential benefits and risks of nanopesticides and contribute to the development of sustainable and responsible strategies for their use.

Introduction

Nanopesticides: A Perspective from Welfare to Risk Assessment is a thought-provoking book that provides a comprehensive overview of the emerging field of nanopesticides. The book takes a unique perspective, focusing on the impact of nanopesticides on human welfare, including their potential benefits and risks. It discusses the benefits of nanopesticides over conventional pesticides, including increased effectiveness, reduced environmental impact, and improved safety.

The book begins by introducing the concept of nanopesticides and their applications in agriculture, highlighting their potential to increase crop yields, reduce environmental impact, and improve food security. It then examines the potential risks associated with the use of nanopesticides, including their impact on human health and the environment, as well as their potential to contribute to the development of pesticide-resistant pests. The book also covers the potential risks associated with nanopesticides, including their toxicity to non-target organisms and the potential for nanoparticles to accumulate in the environment. It explores the various regulatory frameworks that govern the use of nanopesticides in different countries and regions and provides guidance on best practices for their development, testing, and commercialization.

The book goes on to explore the regulatory frameworks that govern the use of nanopesticides and discusses the challenges of developing effective risk assessment strategies for these novel compounds. It provides guidance on best practices for the development, testing, and commercialization of nanopesticides and highlights the need for transparency and communication among stakeholders to ensure responsible use of these technologies.

Overall, *Nanopesticides: A Perspective from Welfare to Risk Assessment* offers a balanced and insightful perspective on this rapidly evolving field. It will be of interest to researchers, regulators, policymakers, and anyone interested in the intersection of nanotechnology and agriculture, and the potential impact of these technologies on human welfare and the environment.

CHAPTER 1

An Introduction to Nanopesticides

VISHWA PATEL,[1] VAIBHAV SHARMA,[2] DARSHINI TRIVEDI,[1] MEENAKSHI SINGH,[3] and K. CHANDRASEKHAR[4]

[1]*Department of Botany, M.S. University of Baroda, Vadodara, Gujarat, India*

[2]*Department of Biotechnology, Himachal Pradesh University, Shimla, Himachal Pradesh, India*

[3]*Department of Ecology and Biodiversity, Sustaina Greens LLP, Vadodara, Gujarat, India*

[4]*Department of Biotechnology, Vignan's Foundation for Science, Technology, and Research, Vadlamudi, Guntur, Andhra Pradesh, India*

ABSTRACT

The development of nanopesticides represents a significant advancement in pest management by enhancing traditional pesticide formulations through nanoparticle delivery systems and active nanoparticles as carrier agents. This innovative approach offers several advantages over conventional methods, including selective toxicity and controlled biodegradability of inorganic nanoparticles. However, the indiscriminate use of nanopesticides may contribute to pesticide resistance in both target and non-target species. Additionally, limited research exists on their environmental fate and potential adverse effects on non-target organisms. To ensure the sustainable and safe implementation of nanopesticides, it is essential to focus on environmentally friendly designs, cost-effective production methods, and rigorous ecotoxicological assessments. Integrating these factors into nanopesticide development

Nanopesticides: A Perspective from Welfare to Risk Assessment.
Garima Awasthi, Payal Lodha, and Gajendra Pal Singh (Eds.)

DOI: 10.1201/9781003605874-1

can help overcome the limitations of conventional pesticides while promoting eco-friendly alternatives with enhanced efficacy and safety. Furthermore, adherence to regulatory guidelines is crucial to minimize potential ecological and human health risks associated with nanopesticide applications.

1.1 NANOPESTICIDES: AN EMERGING TECHNOLOGY

Nanotechnology has played a significant role in nearly all areas of research, including chemistry, ecology, biology, and most notably, biotechnology. The recent fusion of biotechnology and nanotechnology has given rise to a new field, nanobiotechnology, which encompasses various aspects of biotechnology such as research, industry, sewage treatment, and agriculture (Omran, 2020). According to Sun et al. (2020), most applications of nanobiotechnology in agriculture focus on exploring new nanomaterials as sustainable sources of nanopesticides. Certain synthetic polymers, including polyethylene, glycol, and polylactic acid, have demonstrated extensive use in the pharmaceutical and cosmetic industries and show considerable potential as insecticide carriers. However, these chemical pesticides are largely ineffective against *Fusarium* wilt, a disease that affects crops such as tomatoes and lettuce (Yadav et al., 2021b). Moreover, the long-term use of chemical pesticides poses serious environmental risks. Engineered nanomaterials or nanoformulations play an active role in pest management and plant protection. The successful synthesis of nanomaterials from magnesium oxide (MgO), magnesium hydroxide [$Mg(OH)_2$], copper oxide (CuO), and zinc oxide (ZnO) has been achieved using aqueous extracts of green peach aphids and *Chamaemelum nobile* flowers (Raj et al., 2021). Nanopesticides outperform traditional pesticides in terms of adsorption, volatilization, tissue permeability, controlled release, and other properties. This chapter introduces nanopesticides, their commercial formulations, agricultural applications for crop protection, soil-system interactions, and their impact on non-target organisms. Additionally, consumption patterns, environmental implications, and the market potential of nanopesticides are discussed.

1.1.1 CLASSIFICATION OF NANOPESTICIDES

Nanopesticides are a type of nanotechnology-based agricultural product that uses nanoparticles to control pests and diseases in crops.

Nanopesticides can be classified based on their composition, mode of action, and method of delivery. Here are some common classification categories for nanopesticides:

1. **Metal-Based Nanopesticides:** These types of nanopesticides use nanoparticles made of metals such as silver, copper, and zinc. These nanoparticles can be effective in controlling plant pathogens and insect pests.
2. **Carbon-Based Nanopesticides:** Carbon-based nanopesticides use carbon-based nanoparticles such as fullerenes, carbon nanotubes, and graphene to control pests and diseases in crops. These nanoparticles can be used as delivery vehicles for active ingredients, or as active ingredients themselves.
3. **Polymer-Based Nanopesticides:** Polymer-based nanopesticides use polymeric nanoparticles as delivery vehicles for active ingredients. These nanoparticles can be designed to release the active ingredient over time, providing long-lasting pest control.
4. **Lipid-Based Nanopesticides:** Lipid-based nanopesticides use lipid nanoparticles as delivery vehicles for active ingredients. These nanoparticles can be designed to target specific pests or plant diseases and can be used in combination with other control methods.
5. **Mode of Action-Based Nanopesticides:** Nanopesticides can also be classified based on their modes of action, such as those that disrupt insect feeding behaviour or those that interfere with the fungal cell wall.
6. **Delivery-Based Nanopesticides:** Nanopesticides can also be classified based on their delivery methods, such as those that are applied as sprays, those that are applied as seed treatments, or those that are applied through irrigation systems.

It is important to note that the classification of nanopesticides is still evolving as research in this area continues and that the safety and efficacy of these products must be carefully evaluated before they are used in agricultural settings.

Depending on the type of nanomaterial utilised for packaging, nanopesticides can be categorised into organic and inorganic. For instance, when organic nanomaterials are utilised to coat molecules, they are categorised as organic active ingredients with nanocarrier complexes; however, when no nanocarrier is employed in packaging, they are termed as inorganic synthetic active ingredients without carriers, such as Cu $(OH)_2$ (Copper hydroxide).

1.1.1.1 ORGANIC ACTIVE INGREDIENT (AI)-NANOCARRIER COMPLEXES (POLYMERS)

Since organic active ingredients are naturally easily degradable, leave no toxic residue after being degraded by soil microflora, and are also readily available for a lower cost both environmentally and economically, scientists are continually concentrating their efforts on developing efficient new polymeric materials for plant protection. The most researched form of chitin is chitosan, which is a deacetylated form that is found in the cell walls of fungus and insect exoskeletons (Sun et al., 2020). A study by Kashyap et al. (2015) unveiled the biocompatibility of chitosan-controlled release mechanism, which makes them an effective barrier for active compounds, addressing one of the main drawbacks of nanopesticides.

Alginate is incredibly biocompatible, easily degrades, and does not contaminate the environment like chitosan. Alginate has antifungal and antibacterial qualities when mixed with essential oils such as turmeric and lemon grass (Natrajan et al., 2015). The unstable nature of alginate nanocarriers was discovered by Kumar et al. (2015), which resulted in active ingredient leakage; however, the limitations can be easily overcome with better active ingredient release when the nanocarriers were produced by combining two or more polymers, such as chitosan and alginate (Sun et al., 2020).

Cyclodextrins are a by-product of enzymatic processes that break down starch into 6, 7, or 8 subunits of glucose. The truncated cone structure of cyclodextrins enables the development of non-covalent interactions with a number of biological components. Campos et al. (2018) investigated that co-loading carvacrol and linalool in cyclodextrin-chitosan nanocarriers reduced *Tetranychusurticate* oviposition. Moreover, Mishra et al. (2019) confirmed the remarkable antibacterial efficacy of Cyclodextrins against *Streptococcus oralis* and *Streptococcus mutans*.

Starch is a homopolysaccharide that contains many units of glucose branch at 1,6 links and is a promising material in culinary, pharmaceutical, and agriculture-based industries. Starch nanocapsules, avermectin when loaded with the nanocarrier demonstrated a regulated release of avermectin over the course of two weeks (Li et al., 2016). Similarly, Ihegwuagu et al. (2016) reported an increased surface area of nanoparticles by 95–98% by incorporating nanosilver into cassava starch microcapsules, which led to a more sustained release of silver nano-dichlorvos and nano-chlorpyrifos insecticides over a 21-day period.

The most prevalent polymer in nature, cellulose is biodegradable, biocompatible, non-toxic, and least expensive. Ibrahim et al. (2022), revealed growth of *Spodoptera littoralis*, was impacted when citronella essential oil was loaded into a nanocarrier consisting of chitosan nanoparticles and cellulose nanofibers. Likewise, more effectiveness in the eradication of neonicotinoid insects was found using modified cellulose nanoparticles with methacrylic acid (Yuan et al., 2018).

1.1.1.2 INORGANIC ACTIVE INGREDIENTS WITHOUT CARRIERS

As the name implies, inorganic active ingredients are substances that do not include carbon in their molecular structure. These substances are often metals and their metallic oxides. These substances frequently function as potent insecticides on their own, without the aid of nanocarriers. These nanoparticles have a remarkably high proportion of surface area to volume, varied pore sizes, and are exceptionally thermally stable. The present generation of nanoparticles is recognized to eliminate the advantages of conventional techniques, for example, non-target action, restricted solubility, and ecological toxicity (Yadav et al., 2021a). Silver nanoparticles (AgNPs), zinc nanoparticles (ZnNPs), copper and copper oxide nanoparticles (CuNPs and CuONPs) and TiO_2 (Titanium dioxide) nanoparticles are a few of the well-known examples of inorganic nanopesticides.

One of the best nanopesticides that is currently recognized and being explored is silver nanoparticles (AgNPs). They have exceptional antifungal, antibacterial, and pesticidal action. An essential analytic tool for the primary identification of plant diseases by *Anopheles subpictus* and *Culex quinquefasciatus* larvae double when mixed with herbal extracts of plants like *Azadirachta indica* (Kumar et al., 2020). Compared to traditional pesticides, AgNPs offer cheaper manufacturing costs, lower human toxicity, and decreased insect resistance when used at appropriate dose levels (Yadav et al., 2021a).

One of the naturally occurring titanium oxides has little toxicity and very few unintended biological consequences. In many industries, including the food industry, pharmaceuticals, and agriculture, it is widely utilised to inhibit fungal growth. According to a study, cauliflower can be treated with titanium dioxide as a new pesticide to combat the diamond-back moth and *Plutella xylostella* (Preetha et al., 2018). Another recent investigation onTiO_2 was employed to eliminate the cotton leaf worm *Spodoptera littoralis* at a range of concentrations, 1000, 500, 250, 125, 62.5, and 31.25 ppm of TiO_2NPs, the mortality rate was fairly high. When mixed with zinc oxide, titanium dioxide also exhibited insecticidal properties against *Bactericera cockerelli,* and *Fusarium oxysporum* in tomatoes (Javed et al., 2022).

Since, they are easily accessible, inexpensive, and environmentally benign, zinc oxide nanoparticles (ZnONPs) are the best options for nanoparticle formulation in agriculture (Yadav et al., 2021b).ZnONPs exhibit repressive effects on fungi like *Mucor plumbeus, Rhizopus stolonifera, Fusarium oxysporum,* and *Alternaria alternata* (Pho et al., 2020).

1.1.2 COMMERCIAL FORMULATIONS OF NANOPESTICIDES

The development of several commercial nanoformulations for crop protection has resulted from research in nanotechnology. Depending on their intended use, nanopesticides are designed to increase solubility, limit the discharge of active chemicals, stop deterioration, and do other things. In order to achieve these aims, the chemical nature carrier molecule has undergone chemical changes, classified as polymeric/lipid/bimetallic/clay-based nanomaterials, and so forth.

1.1.2.1 NANOEMULSIONS

Nanoemulsions are a type of emulsion with droplet sizes typically in the range of 20-200 nanometers. They are thermodynamically stable, transparent or translucent, and can be composed of oil-in-water (O/W) or water-in-oil (W/O) phases. Nanoemulsions have a high surface area to volume ratio, which makes them useful in a variety of applications, including food, pharmaceuticals, and cosmetics. One of the key benefits of nanoemulsions is their ability to increase the bioavailability of poorly water-soluble drugs and nutrients. The small droplet size of nanoemulsions allows them to be more easily absorbed by the body, which can improve the efficacy of the active ingredients. Nanoemulsions can also be used to improve the texture, stability, and shelf life of food and cosmetic products. Nanoemulsions are typically formed using high-pressure homogenization or microfluidic techniques. In high-pressure homogenization, the emulsion is forced through a small nozzle under high pressure, which breaks up the droplets into smaller sizes. In microfluidic techniques, the emulsion is formed by mixing the phases in a microfluidic channel, which allows for precise control over the droplet size and composition. Despite their many potential benefits, there are also concerns about the safety of nanoemulsions, particularly in food and cosmetic applications. The small size of the droplets in nanoemulsions can potentially lead to increased absorption of certain ingredients, which may have unintended health consequences. As with any new technology, the safety and efficacy of nanoemulsions must be carefully evaluated before they are widely adopted.

The chemical's active ingredient is released as nanosized droplets in water as an oil-in-water (O/W) emulsion, with surfactant molecules constrained at the pesticide-water interface. The delivery strategy for pharmaceutical compounds is a nanoemulsion, a colloidal particle system of a submicron size range. Their sizes range from 10 to 1,000 nm. These carriers have a negative charge and an amorphous, lipophilic surface. There are three types of nanoemulsions: (a) water-in-oil, (b) oil-in-water, and (c) bi-continuous. The emulsifier assists in stabilizing nano emulsions through steric hindrance and repelling electrostatic interactions. In addition to drug delivery, where O/W nanoemulsions are used to transport hydrophobic medicines, flavored nanoemulsions with better digestibility, and cosmetics for various skin applications.

1.1.2.2 NANOCAPSULES

Nanocapsules are a type of nanoscale particle that can be used to encapsulate and deliver a variety of active ingredients, including drugs, nutrients, and other bioactive compounds. They consist of a shell made of a biocompatible or biodegradable polymer, lipid, or silica, which surrounds a core containing the active ingredient. Nanocapsules can be designed to release their payload in a controlled manner, based on external or internal stimuli such as temperature, pH, or enzymatic activity. This allows for targeted and sustained release of the active ingredient, which can improve its efficacy and reduce side effects. Nanocapsules can be synthesized using a variety of techniques, including emulsion polymerization, solvent evaporation, and layer-by-layer assembly. The choice of technique depends on the properties of the active ingredient and the desired release profile. Nanocapsules have numerous applications in medicine, including drug delivery, gene therapy, and imaging. They can be used to target specific tissues or cells in the body, reducing the amount of drug needed and minimizing off-target effects. Nanocapsules can also improve the bioavailability of poorly soluble drugs, prolong their circulation time in the body, and protect them from degradation. Nanocapsules also have potential applications in the food and cosmetic industries. They can be used to encapsulate flavors, fragrances, and other active ingredients, improving their stability and allowing for controlled release. In cosmetics, nanocapsules can be used to deliver anti-ageing compounds, vitamins, and other ingredients to the skin.

A protective matrix "shell" and active components "core" make up nanocapsules, with a therapeutic component. Drug-filled nanocapsules may include cell surface receptors on their surface that bind to cancer cells or other target cells and release a biological material when they come into touch with that particular tissue. From 10 nm to 1000 nm, polymeric nanocapsules come in a range of sizes and forms. A liquid or solid core with a cavity where the medicine is put is surrounded by a special polymer membrane consisting of organic or synthetic polymers. Because of the protective coating, which is typically pyrophoric, quickly oxidized, and prevents the release of active chemicals, they have garnered a lot of attention (Diaspro et al., 2002).

The biomedical industry continues to be interested in nanocapsules because, among other things, they may be employed for the restricted delivery and targeting of medications against the protection of enzymes, proteins, and alien cells.

1.1.2.3 NANOPARTICLES – METALS, METAL OXIDES, AND NANOCLAYS

Metallic nanoparticles (MNPs) are frequently coated with organic or inorganic materials and are used in a variety of consumer products, including lotions, shampoos, clothing, footwear, and plastic containers.

Their use has increased significantly with the advent of pilot-scale MNP manufacturing utilizing new, cost-effective methods (Diegoli et al., 2008). Among the most popular metal-oxide nanoparticles, zinc oxide (ZnO), titanium dioxide (TiO_2), and aluminum oxide (Al_2O_3) derivatives are the most widely used commercial materials in research due to their variations in photolytic capabilities.

Nanoclay is a type of clay that has been processed into tiny particles, typically with dimensions of less than 100 nanometers. Due to its unique properties, nanoclay has become a popular material for use in composites, particularly in industries such as automotive, aerospace, and construction. One of the key advantages of using nanoclay in composites is its stability, which means that it is not easily affected by changes in temperature or humidity. Additionally, nanoclay has a high swelling capacity, which allows it to absorb large amounts of liquid without losing its structural integrity. The layered crystalline silicate nanoparticles known as nanoclay have a variety of unusual characteristics and offer plenty of room for nanotechnology experimentation. Nanoclay imparts superior properties to composites by virtue of its stability, swelling capacity, interlayer spacing, enhanced hydration, and robust chemical reactivity. These qualities allow nanoclay to improve the features of composites. The many different varieties of nanoclays may be categorized as montmorillonite, bentonite, kaolinite, hectorite, and halloysite. These classifications are based on the chemical compositions and morphologies of the nanoclays (Sillu et al., 2021).

1.1.2.4 NANOFIBRES

Nanofibers are ultrafine fibers with diameters in the range of nanometers (typically less than 100 nm). They are often produced by electrospinning, a process that uses an electric field to create fibers from a liquid or polymer solution. Nanofibers have a very high surface area to volume ratio, which makes them useful in a wide range of applications. One of the most promising applications of nanofibers is in the field of tissue engineering, where they can be used as scaffolds for growing new tissues and organs. Nanofibers can also be used in filtration, drug delivery, and energy storage. Nanofibers have unique mechanical, electrical, and chemical properties that make them attractive for use in a variety of fields. For example, they can be used to make lightweight, high-strength materials for aerospace and automotive applications. They can also be used in electronics and photonics to make faster and more efficient devices. Despite their many potential applications, the production of nanofibers is still relatively expensive and time-consuming, and there are still many challenges to be overcome before they can be widely used in industry. However, research in this area is ongoing, and it is likely that nanofibers will play an increasingly important role in many different fields in the coming years.

As the fiber diameter approaches the nanoscale, nanofibers offer small hole sizes and enormous surface areas with innovative and greatly improved physical, chemical, and biological properties. Therefore, gained prominent exposure in the recent past. Polymer nanofibers have been created using a variety of methods. One of the novel methods is polystyrene-based electrospinning, where nanofibers with ultimate dimensions are equal to those of single-wall carbon nanotubes and showed greater pesticide adsorption capacity (Ruggieri et al., 2015). According to Barbara et al. (2019), the easiest method for localised pesticide distribution involves covering Soyabean seedlings with electrospun cellulose diacetate (CDA) nanofibers that contain certain targeted chemicals against nematodes and fungus growth. Due to CDA's favourable electro-spinnability, low water solubility, and potential biodegradability, it is utilised as the preferred polymer with no detrimental impacts on seed germination.

1.1.2.5 NANOGELS

Nanogels are a type of nanoscale hydrogel, which are three-dimensional networks of polymers that are capable of absorbing and retaining large amounts of water or other fluids. They are typically in the range of 10 to 100 nanometers in size and have unique properties that make them useful in a variety of applications. Nanogels are typically made from biocompatible or biodegradable polymers, such as polyethylene glycol (PEG), polyvinyl alcohol (PVA), or chitosan. They can be designed to respond to changes in temperature, pH, or other environmental factors, which allows them to release drugs or other active compounds in a controlled manner. Nanogels can be synthesized using a variety of techniques, including emulsion polymerization, microfluidics, and self-assembly. However, there are still many challenges to be overcome in the development and production of nanogels, including improving their stability and biocompatibility, as well as optimizing their drug-release properties.

Overall, nanogels hold great promise for a wide range of applications in medicine, biotechnology, and materials science, and research in this area is ongoing. Nanogels are 3D hydrogel substances that have a unique high water-retention capacity without dispersing in the liquid because of crosslinked swellable polymer networks. The physical properties have a huge capacity for guest molecules, more biocompatible in foliar wettability, photostability, and precision targeting (Shen et al., 2022). Nanogels actively participate in the delivery process to generate a controlled, triggered response at the target spot while shielding the payload from deterioration. Moreover, minimising pollution and pesticide exposure is made possible by nanogels' tendency to degrade. The nanocarriers for delivering pesticides are promising because of this system's easy setup, good qualities, and environmental friendliness (Luo et al., 2021).

1.2 PERSPECTIVES OF NANOPESTICIDES IN AGRICULTURE

Sustainable agriculture must be seen as an ecosystem approach that promotes the peaceful coexistence of biotic and abiotic creatures, as well as synchronized food chains and related energy balances. Technological advancement regarding nanochemicals to increase agricultural productivity is approved by governmental laws and also implemented in the food industry (Chaud et al., 2021; Kannan et al., 2023). Thus, nanotechnology is the cutting-edge technology in the global agricultural sector's food supply chain, including crop production, use of agrochemicals, and precision farming techniques in agricultural nanotechnology may need to pay greater attention to several specific areas:

- New delivery methods that are safe and environmentally friendly for conveying plant nutrients, specialty food/feed ingredients, and so forth. These systems may potentially be used in pharmaceutical applications.
- Nanotechnology-related (bio)sensors in agricultural production and pest management. Through intelligent food packaging combined with nanosensors, consumers may always obtain accurate information on the condition of a specific food product.
- A detailed analysis of the nanomaterials' life cycle evaluation, including measurements of their size, dose, exposure time, surface chemistry, immunological response, accumulation, retention time, and so forth. To recognize, validate, and access the effects of each

nanomaterial on whole ecosystems, new analytical techniques must be developed. A complete databank has to be improved, and there needs to be global collaboration on concepts, rules, and regulations in order to manipulate this information. Also, the government should provide guidelines and roadmaps for reducing the risks connected with employing nanotechnological products.

- With the involvement of several stakeholders, new dialogues should enable the consequences of nanotechnology on human life, environment, and financial decisions.
- Long-term, nanotechnology may offer creative and cost-effective development avenues for human nutrition across the globe.

1.2.1 CROP PROTECTION

Nanopesticide formulations have the potential to revolutionize crop management by enhancing hydration properties, bioavailability, and agro-chemical protection from environmental damage. Yet, the features of the nanomaterials are also on the edge of being cytotoxic and genotoxic. Figure 1.1 depicts two ways of nanopesticide formulation delivery in agriculture.

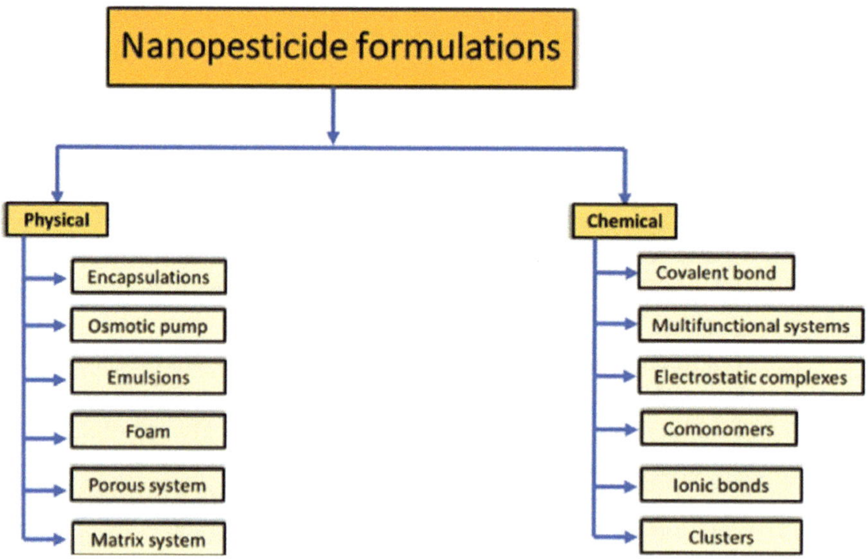

FIGURE 1.1 Nanopesticides controlled delivery mechanism involves two approaches: physical and chemical, where the formulation is released as carrier agents.

1.2.2 INTERACTION WITH SOIL SYSTEMS

Nanopesticides from agricultural and industrial wastewater runoff reach the water supply when soil penetration leaching happens during a precipitation event, altering its quality, extending human exposure duration, and creating environmental problems (Bombo et al., 2019). The toxic effects of nanoparticles are because of biomimetic properties, high dissemination capacity, and bioaccumulation ability in food and water environments, as well as in all organisms, notably mammals. When it comes to humans, the spectrum of harmful effects linked to individual susceptibility and the length of time spent in contact with nanoparticles results in acute and chronic pathological manifestations either immediately after exposure or years later as a result of bioaccumulation and the unique properties of nanoparticles.

Côa et al. (2020) suggested that through tailored therapies to reduce crop losses brought on by abiotic and biotic stresses, nanotechnology intends to shift agriculture from a time of reckless resource application and environmental damage to a new era of sophisticated systems with improved material use efficiency. Pesticides are necessary for agriculture to control plant diseases and insect pests, without compromising on nature. Yet, various environmental and health problems as a result of the rising per-hectare concentration of these chemicals. According to Hayles et al. (2017), the nanotechnological revolution has led to the development of more powerful nanopesticides as a solution to the issues with traditional pesticides. The main benefit comes from the particles' small size, which aids in even dispersion of the components on insect surfaces and produces a better effect than traditional pesticides. Due to their better efficacy and lower dosage requirements, nanoparticles are increasingly being used as nanopesticides, nanofertilizers, and nano-delivery systems. However, during or after the application, living beings are exposed to nano-entities with mostly unknown histories of their interaction with biological systems. Therefore, for a sustainable transition, a deeper knowledge of their interactions and potential negative effects is also essential before their widespread use in crop production and crop protection (Huang et al., 2018).

In addition, the harmful pesticide residuals left on earth may prove risky to the non-parasitic and advantageous soil microbiota by eliminating their species composition. Pesticides, especially those that are saprophytic and spore-forming, can limit the growth of some microbial populations while supporting the growth of others. Pesticide residues and their metabolites can pile up in the soil, changing its chemical makeup by decomposing the microbial population. This results in the lowering of soil potency and its capacity to maintain life, disturbing the soil microenvironment (Zielińska et al., 2020).

1.2.3 IMPACT ON NON-TARGETED ORGANISMS

Pesticides can have unintended effects on non-target organisms, particularly important soil invertebrates such as earthworms, predatory mites, centipedes, and carabid beetles. These organisms play crucial roles in soil health and ecosystem functioning, and their loss or decline can have far-reaching impacts. Pesticide residues can also accumulate in the bodies of organisms higher up the food chain, leading to periods of toxicity that can have cascading effects on entire ecosystems.

Chlorinated hydrocarbons are particularly problematic in this regard, as they are among the most ecotoxic chemicals in use. To mitigate the unintended effects of pesticides, it is important to apply them sparingly and with great care. In cases where pesticides are necessary, it is important to use methods that minimize their impact on non-target organisms. For example, studies of accidental bycatch during fishing expeditions can help identify alternative methods for controlling target organisms without harming non-target species. In the case of tropical tuna fishing, for instance, many species besides tuna, such as other bony fishes, sharks, rays, and turtles, are frequently caught, some of which may be deliberately targeted. Understanding the impacts of fishing on these non-target species is crucial for developing more sustainable and environmentally friendly fishing practices (Chaud et al., 2021).

1.3 RISK ASSESSMENT OF NANOPESTICIDES

Nanopesticides, like any other nanotechnology-based products, pose potential risks to human health and the environment. Therefore, it is essential to conduct a thorough risk assessment of these products before they are approved for commercial use. A risk assessment of nanopesticides typically includes an evaluation of their toxicity, exposure, and hazard potential. The toxicity assessment of nanopesticides involves determining the adverse effects that these products may have on living organisms.

This includes evaluating their acute and chronic toxicity, as well as their potential to accumulate in organisms over time. Additionally, the exposure assessment involves determining how people and the environment may come into contact with the nanopesticides. This includes evaluating the risks posed by their production, use, and disposal. The hazard potential of nanopesticides is determined by evaluating their physicochemical properties, including their size, shape, surface charge, and reactivity. These properties can affect how the nanopesticides interact with biological systems and the environment, potentially leading to unintended effects.

Risk assessment of nanopesticides is a complex process that requires input from multiple disciplines, including toxicology, epidemiology, environmental science, and engineering. There is currently a lack of standardized methods for assessing the risks of nanotechnology-based products, including nanopesticides. Therefore, developing such methods is essential to ensure that the potential risks of these products are adequately evaluated. In conclusion, risk assessment of nanopesticides is essential to ensure that these products are safe for human health and the environment. The complexity of these products and their potential risks require a multidisciplinary approach to risk assessment. Standardizing risk assessment methods is necessary to ensure that the potential risks of nanopesticides are adequately evaluated and managed.

It is crucial to keep an eye on the non-target consequences of these nanomaterials on the environment, other living things, and ecosystems as their use spreads to every industry, including the food, pharmaceutical, water treatment, and agricultural sectors. Agriculture has grown to be one of the main pillars of our economy, and diseases, whether they take the shape of insects, fungi, or bacteria, harm produce most frequently each year. In India, pests eat about 15% to 25% of the food produced each year (WHO, 2019). Hence, the use of pesticides in agriculture, whether organic or inorganic, has become quite prevalent. More recently, nanopesticides have shown great promise as a solution to the drawbacks of chemical-based pesticides (Kannan et al., 2023).

1.3.1 TOXICITY OF NANOFORMULATIONS

A study by Chaud et al. (2019), revealed that materials used in the production of a certain nanoparticle may induce a toxic effect not only on the plant itself but also on the humans that may consume the plant as food.

The effect of clomazone and nanoparticles was studied by exposing the tadpoles *(Lithobates catesbeianus)* to the mixture and it was found that it caused issues with the development of reproductive organs in the tadpoles.

Another study conducted on Enchytraeids revealed that prolonged exposure to atrazine-based nanoparticles 20% mortality rates in adult enchytraeids whereas the mortality rate in juvenile enchytraeids was more than 50%. In terms of egg hatching in enchytraeids, a substantial decline in number was observed that suggests atrazine at 200 mg/kg hampers the reproductive cycle (Gomes et al., 2019). The harmful effects of nanopesticides have also been observed in plants. For instance, it was reported that atrazine-loaded poly(epsilon-caprolactone) nanoparticles caused lipid peroxidation in the leaves of *Zea mays*, $Cu(OH_2)$-based nanoparticles caused chlorophyll degradation in maize plants, and *Brassica napa* (mustard) showed early signs of leaf chlorosis when exposed to diuron nanoformulation. Atrazine, a chemical that is frequently used to protect maize and mustard crops, has been demonstrated to have negative effects on aquatic life and the quality of drinking water.

A study by Demir, 2020 to test the toxicity of silver and sulfur nanoparticles was conducted on *Drosophila* and reported considerably shorter lifespans and a higher fatality rate for adult flies. *Drosophila* toxicity is a new field of study wherein *Drosophila* is exposed to different nanoparticles through food, injection, or aerosols in a controlled setting and is then observed for a number of parameters like DNA damage, morphological flaws, and neurological health. The US Federal Insecticide, Fungicide, and Rodenticide Act's Scientific Advisory Panel, however, stated that the health and environmental risks of nanopesticides may not be adequately predicted or addressed by present techniques. Europe practices stringent regulatory frameworks and authorization for the assessment of human health as well as environmental hazards of nanopesticides before being released in the market. Recently, India also evaluates the risk assessment of nano-agriproducts as per new guidelines published regarding defining nanopesticides and nano fertilizers (Grillo et al., 2021).

1.3.2 ENVIRONMENTAL REPERCUSSIONS

It is challenging to research the toxicity that nanoformulations can produce in the biosphere due to the wide range of environmental interactions that might occur, and only a small number of the components that cause the accumulation and toxicity of the nanoformulations have been inadequately investigated so far. The alteration of nanomaterials may be one such element; typically, photooxidation and photoreduction of nanomaterials are brought about by sunshine. It is also possible for nanomaterials to aggregate in homogeneous or heterogeneous ways, which causes sedimentation and lengthens their life in the environment. Combining organic and inorganic ligands with nanomaterials can occasionally have unanticipated harmful consequences (Côa et al., 2020).

According to a study, 260,000–309,000 tonnes of nanomaterials were disseminated into the environment, with 63–91% of that amount ending up in landfills, 8%–28% in soils, 0.4–7% through water, and 0.1–1.5% into the air (Sohn et al., 2015). It is currently unknown whether sewage treatment plants can sufficiently remove nanoparticles from the water so that there is no bioaccumulation and toxic build-up. According to toxicological studies, freshwater life forms including fish, algae, and crabs are extremely harmful to nanomaterial oxides like TiO_2. Nanometal oxides can also build in the soil, where they can be absorbed by plants and cause harm to other organisms as well (Gaur et al., 2020; Grillo et al., 2021). The effects of nanopesticides and nanomaterials on the environment are therefore mediated through nano-bio-eco-interactions, which are controlled by biological, chemical, physical, and geological processes that can change the physicochemical characteristics of nanoparticles.

1.4 SCOPE OF THE NANOPESTICIDE MARKET

On a global scale, the nanopesticides market is forecasted to achieve USD 1.9 billion with a listed CAGR of 13.66% by 2030. The Covid pandemic has badly impacted the food and agriculture industries and people's livelihoods. Therefore, nanopesticides resume a hope of protecting plants from pest control, the longevity of crops against environmental stress and minimising crop degradation by plant pathogens. Moreover, nanopesticides boost the freshness of perishable foods and help in nutritional retention for a longer period.

According to Figure 1.2, Europe is leading the nanopesticide market as per the revenue collection in the year 2021, followed by Asia Pacific and North America. The top players in this segment are Marrone Bio

Innovations (United States), Bioworks (United States), Valent Biosciences (United States), Andermatt Biocontrol (Switzerland), Stockton (Israel), Bayer (Germany), Camson Bio Technologies (India), Cortevaagriscience (United States), and so forth. The innovative agricultural solutions enabled intensive farming and automation to maximize food production, thereby creating a vast space for pesticide consumption. As a result, countries are promoting nanopesticide use in lieu of general guidelines for acceptable chemical formulations. For instance, a Pesticide Data Program (PDP) devised by the Agriculture Department in the U.S.A., monitor pesticide usage across the nation and develop a comprehensive database on pesticide consumption and its residue. In a similar manner, an Integrated Pest Management (IPM) programme is implemented and practiced in India to strengthen the fundamentals of plant protection while promoting agricultural productivity. This was done in order to safeguard plants from pests. These techniques contribute to satisfying the growing demand for food from the population as a whole and helping to avert food shortages.

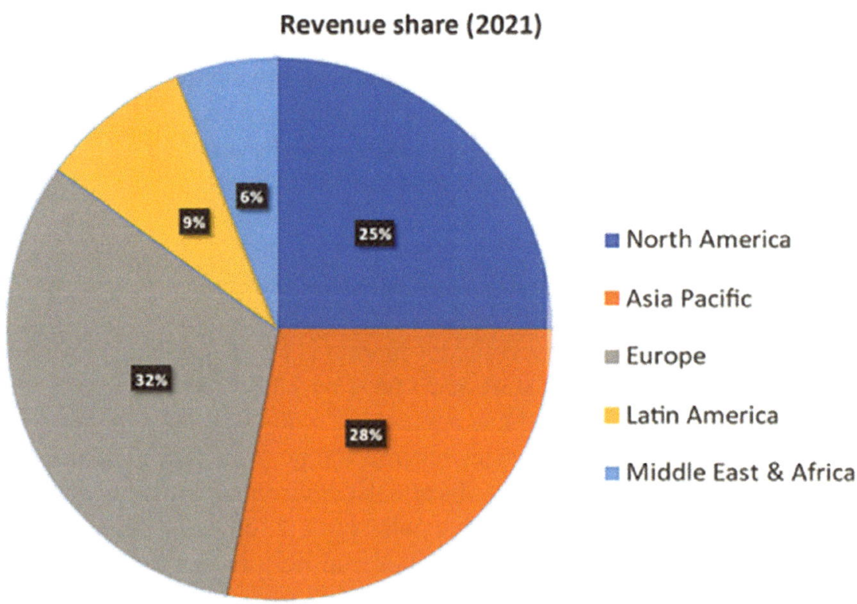

FIGURE 1.2 The region-wise market size of nanopesticide (2021).

Source: Adapted from https://www.precedenceresearch.com/nanopesticides-market.

1.5 CONCLUSION

An excellent method of managing pests is made possible by the innovative improvement of traditional pesticide formulations, nanoparticle delivery systems, and active nanoparticles as carrier agents. Contrary to conventional pest management methods, this more environmentally friendly approach has a number of advantages, like selective toxicity and limited biodegradability of inorganic nanoparticles, and its random use may result in pesticide resistance in both target and non-target species. However, there is a dearth of knowledge and scanty data about these nanoparticles' environmental fates and their possible damaging impacts on non-target animals. So, greater attention must be paid to the creation of environmentally friendly nanopesticides as well as the predictable negative impacts on the environment, non-target animals, and people. Such elements, if incorporated in the chemical design of nanopesticides may overcome the drawbacks of conventional preparations, sustainable promotion of eco-friendly chemicals, low-cost production strategies, and practical utility comparison with traditional analogues. Moreover, the safe introduction of nanopesticides requires stringent ecotoxicological assessment and other measures as suggested by authorized regulatory bodies.

KEYWORDS

- **ecotoxicological**
- **nanocapsules**
- **nanoemulsions**
- **nanofibres**
- **nanogels**
- **nanopesticides**

REFERENCES

Barbara, V. F., Pirzada, T., Mathew, R., Sit, T. L., Opperman, C., & Khan, S. A. (2019). Electrospun polymer nanofibers as seed coatings for crop protection. *ACS Sustainable Chemistry & Engineering, 7*(24), 19848–19856.

Bombo, A. B., Pereira, A. E. S., Lusa, M. G., De Medeiros Oliveira, E., De Oliveira, J. L., Campos, E. V. R., De Jesus, M. B., Oliveira, H. C., Fraceto, L. F., & Mayer, J. L. S. (2019). A mechanistic view of interactions of a nanoherbicide with target organisms. *Journal of Agricultural and Food Chemistry, 67*, 4453–4462.

Campos, E. V., Proença, P. L., Oliveira, J. L., Pereira, A. E., de Morais Ribeiro, L. N., Fernandes, F. O., & Fraceto, L. F. (2018). Carvacrol and linalool co-loaded in β-cyclodextrin-grafted chitosan nanoparticles as sustainable biopesticide aiming pest control. *Scientific Reports, 8*(1), 7623.

Chaud, M., Souto, E. B., Zielinska, A., Severino, P., Batain, F., Oliveira-Junior, J., & Alves, T. (2021). Nanopesticides in agriculture: Benefits and challenges in agricultural productivity, toxicological risks to human health, and the environment. *Toxics, 9*(6), 131.

Côa, F., Bortolozzo, L. S., Petry, R., Da Silva, G. H., Martins, C. H., de Medeiros, A. M., & Martinez, D. S. T. (2020). Environmental toxicity of nanopesticides against non-target organisms: The state of the art. In *Nanopesticides: From Research and Development to Mechanisms of Action and Sustainable Use in Agriculture* (pp. 227–279). Springer.

Demir, E. (2020). Drosophila as a model for assessing nanopesticide toxicity. *Nanotoxicology, 14*(9), 1271–1279.

Diaspro, A., Krol, S., Cavalleri, O., Silvano, D., & Gliozzi, A. (2002). Microscopical characterization of nanocapsules templated on ionic crystals and biological cells toward biomedical applications. *IEEE Transactions on Nanobioscience, 1*(3), 110–115.

Diegoli, S., Manciulea, A. L., Begum, S., Jones, I. P., Lead, J. R., & Preece, J. A. (2008). Interaction between manufactured gold nanoparticles and naturally occurring organic macromolecules. *Science of the Total Environment, 402*, 51–61.

Gaur, N., Sharma, N., Dahiya, A., Yadav, P., Ojha, H., Goyal, R. K., & Sharma, R. K. (2020). Toxicity and regulatory concerns for nanoformulations in medicine. In *The ELSI Handbook of Nanotechnology: Risk, Safety, ELSI and Commercialization* (pp. 333–357). Springer.

Gomes, S. I., Scott-Fordsmand, J. J., Campos, E. V., Grillo, R., Fraceto, L. F., & Amorim, M. J. (2019). On the safety of nanoformulations to non-target soil invertebrates: An atrazine case study. *Environmental Science: Nano, 6*(6), 1950–1958.

Grillo, R., Fraceto, L. F., Amorim, M. J., Scott-Fordsmand, J. J., Schoonjans, R., & Chaudhry, Q. (2021). Ecotoxicological and regulatory aspects of environmental sustainability of nanopesticides. *Journal of Hazardous Materials, 404*, 124148.

Hayles, J., Johnson, L., Worthley, C., & Losic, D. (2017). *Nanopesticides: A Review of Current Research and Perspectives*. Elsevier Inc. ISBN 9780128042991.

Huang, B., Chen, F., Shen, Y., Qian, K., Wang, Y., Sun, C., Zhao, X., Cui, B., Gao, F., Zeng, Z., et al. (2018). Advances in targeted pesticides with environmentally responsive controlled release by nanotechnology. *Nanomaterials, 8*, 102.

Ibrahim, S. S., Abou-Elseoud, W. S., Elbehery, H. H., & Hassan, M. L. (2022). Chitosan-cellulose nanoencapsulation systems for enhancing the insecticidal activity of citronella essential oil against the cotton leafworm *Spodoptera littoralis*. *Industrial Crops and Products, 184*, 115089.

Ihegwuagu, N. E., Sha'Ato, R., Tor-Anyiin, T. A., Nnamonu, L. A., Buekes, P., Sone, B., & Maaza, M. (2016). Facile formulation of starch-silver-nanoparticle encapsulated dichlorvos and chlorpyrifos for enhanced insecticide delivery. *New Journal of Chemistry, 40*(2), 1777–1784.

Javed, R., Ain, N. U., Gul, A., Arslan Ahmad, M., Guo, W., Ao, Q., & Tian, S. (2022). Diverse biotechnological applications of multifunctional titanium dioxide nanoparticles: An up-to-date review. *IET Nanobiotechnology, 16*(5), 171–189.

Kannan, M., Bojan, N., Swaminathan, J., et al. (2023). Nanopesticides in agricultural pest management and their environmental risks: A review. *International Journal of Environmental Science and Technology*. https://doi.org/10.1007/s13762-023-04795-y.

Kashyap, P. L., Xiang, X., & Heiden, P. (2015). Chitosan nanoparticle-based delivery systems for sustainable agriculture. *International Journal of Biological Macromolecules, 77*, 36–51.

Khan, I., Saeed, K., & Khan, I. (2019). Nanoparticles: Properties, applications, and toxicities. *Arabian Journal of Chemistry, 12*, 908–931.

Kumar, D., Kumar, P., Singh, H., & Agrawal, V. (2020). Biocontrol of mosquito vectors through herbal-derived silver nanoparticles: Prospects and challenges. *Environmental Science and Pollution Research, 27*, 25987–26024.

Kumar, S., Chauhan, N., Gopal, M., Kumar, R., & Dilbaghi, N. (2015). Development and evaluation of alginate–Chitosan nanocapsules for controlled release of acetamiprid. *International Journal of Biological Macromolecules, 81*, 631–637.

Li, D., Liu, B., Yang, F., Wang, X., Shen, H., & Wu, D. (2016). Preparation of uniform starch microcapsules by premix membrane emulsion for controlled release of avermectin. *Carbohydrate Polymers, 136*, 341–349.

Luo, J., Gao, Y., Liu, Y., Huang, X., Zhang, D., Cao, H., Jing, T., Liu, F., & Li, B. (2021). Self-assembled degradable nanogels provide foliar affinity and pinning for pesticide delivery by flexibility and adhesiveness adjustment. *ACS Nano, 15*(9), 14598–14609.

Mishra, P., Seenivasan, R., Mukherjee, A., & Chandrasekaran, N. (2019). Polymer/layered silicate nanocomposites as matrix for bioinsecticide formulations. In *Nano-Biopesticides: Today and Future Perspectives* (pp. 161–178).

Natrajan, D., Srinivasan, S., Sundar, K., & Ravindran, A. (2015). Formulation of essential oil-loaded chitosan–alginate nanocapsules. *Journal of Food and Drug Analysis, 23*(3), 560–568.

Omran, B. A. (2020). Fundamentals of nanotechnology and nanobiotechnology. In *Nanobiotechnology: A Multidisciplinary Field of Science. Nanotechnology in the Life Sciences* (pp. 1–24). Springer. https://doi.org/10.1007/978-3-030-46071-6_1.

Pho, Q. H., Losic, D., Ostrikov, K. K., Tran, N. N., & Hessel, V. (2020). Perspectives on plasma-assisted synthesis of N-doped nanoparticles as nanopesticides for pest control in crops. *Reaction Chemistry & Engineering, 5*(8), 1374–1396.

Preetha, S., Kannan, M., Lokesh, S., Viji, N., Prithiva, J. N., & Gowtham, V. (2018). Titanium dioxide (TiO_2) nanoparticles as a novel insecticide against diamondback moth, *Plutella xylostella*, in cauliflower. *Trends in Biosciences, 11*(21), 2999–3003.

Raj, S. N., Anooj, E. S., Rajendran, K., & Vallinayagam, S. (2021). A comprehensive review on regulatory invention of nanopesticides in agricultural nano formulation and food system. *Journal of Molecular Structure, 1239*, 130517.

Ruggieri, F., D'Archivio, A. A., Di Camillo, D., Lozzi, L., Maggi, M. A., Mercorio, R., & Santucci, S. (2015). Development of molecularly imprinted polymeric nanofibers by electrospinning and applications to pesticide adsorption. *Journal of Separation Science, 38*(8), 1402–1410.

Shen, Y., An, C., Jiang, J., Huang, B., Li, N., Sun, C., Wang, C., Zhan, S., Li, X., Gao, F., Zhao, X., Cui, H., Gooneratne, R., & Wang, Y. (2022). Temperature-dependent nanogel for pesticide smart delivery with improved foliar dispersion and bioactivity for efficient control of multiple pests. *ACS Nano, 16*(12), 20622–20632. https://doi.org/10.1021/acsnano.2c06773.

Sillu, D., & Kaushik, Y. (2021). Immobilization of enzymes onto silica-based nanomaterials for bioprocess applications. In A. Tripathi & J. S. Melo (Eds.), *Immobilization Strategies* (pp. 193–210). Springer. https://doi.org/10.1007/978-981-15-7998-1_11.

Sohn, E. K., Johari, S. A., Kim, T. G., Kim, J. K., Kim, E., Lee, J. H., Chung, Y. S., & Yu, I. J. (2015). Aquatic toxicity comparison of silver nanoparticles and silver nanowires. *BioMed Research International, 2015*, 893049. https://doi.org/10.1155/2015/893049.

Sun, C., Zeng, Z., Cui, H., & Verheggen, F. (2020). Polymer-based nanoinsecticides: Current developments, environmental risks and future challenges: A review. *Biotechnologie, Agronomie, Société et Environnement, 24*, 1–18.

World Health Organization & Food and Agriculture Organization of the United Nations. (2019). *Global situation of pesticide management in agriculture and public health: Report of a 2018 WHO-FAO survey*. World Health Organization. https://apps.who.int/iris/handle/10665/329971 (accessed on 3 October 2024).

Yadav, D. R., Adhikari, M., Kim, S. W., Kim, H. S., & Lee, Y. S. (2021b). Suppression of fusarium wilt caused by *Fusarium oxysporum* f. sp. *lactucae* and growth promotion on lettuce using bacterial isolates. *Journal of Microbiology and Biotechnology, 31*(9), 1241–1255.

Yadav, J., Jasrotia, P., Kashyap, P. L., Bhardwaj, A. K., Kumar, S., Singh, M., & Singh, G. P. (2021a). Nanopesticides: Current status and scope for their application in agriculture. *Plant Protection Science, 58*(1), 1–17.

Yuan, M., Liu, X., Li, C., Yu, J., Zhang, B., & Ma, Y. (2019). A higher efficiency removal of neonicotinoid insecticides by modified cellulose-based complex particle. *International Journal of Biological Macromolecules, 126*, 857–866.

Zielińska, A., Costa, B., Ferreira, M. V., Miguéis, D., Louros, J. M. S., Durazzo, A., Lucarini, M., Eder, P., Chaud, M. V., Morsink, M., & Poinern, G. E. J. (2020). Nanotoxicology and nanosafety: Safety-by-design and testing at a glance. *International Journal of Environmental Research and Public Health, 17*(13), 4657.

CHAPTER 2

Destiny of Nanopesticides in Soil and Water

SUPRIYA KUMARI,[1] SURBHI KUMARI,[1] KUMUD KANT AWASTHI,[2] ANJALI AWASTHI,[3] and RAJBALA VERMA[3]

[1]*Department of Forensic Science, Vivekananda Global University, Jaipur, Rajasthan, India*

[2]*K. R. Mangalam University, Gurugram, Haryana, India*

[3]*Department of Zoology, University of Rajasthan, Jaipur, Rajasthan, India*

ABSTRACT

In order to maximize the potent effects of active substances, nanopesticides (NP) are nanostructures with two to three dimensions and a size between 1 and 200 nm (AI). When opposed to free insecticides, adding AI to nanoparticles has advantages because of their special features. The huge advantages of nanopesticides in agriculture include replacing synthetic fertilizers and pesticides, which increases crop yield. There is, however, a lot of study being done in this area, and it can be difficult to reach farmers because of things like higher production costs, a lack of understanding, effects on people and the environment, and so forth. There are many different items covered by the phrase "nanopesticide," thus it cannot be regarded as standing for just one type of product. There are numerous nanoformulations that blend various surfactants, polymers, and approximately nanometer-sized metal particles. There are several ways

Nanopesticides: A Perspective from Welfare to Risk Assessment.
Garima Awasthi, Payal Lodha, and Gajendra Pal Singh (Eds.)

DOI: 10.1201/9781003605874-2

to make nanoparticles, including chemical synthesis, arthropod-based synthesis, and green synthesis. Pros and cons of these nanopesticides are by no means neglectable. Studies are fewer than needed to comprehensively discuss nanopesticides. Here we are discussing about the destiny of nanopesticides when they are introduced in soil and water. We have discussed the mechanism through which nanopesticides are allowed to enter in water and soil and what happens after that like sorption, degradation, and so forth (Figure 2.1).

FIGURE 2.1 Graphical abstract.

2.1 INTRODUCTION

Nanopesticides are made up of either incredibly tiny adjuvant pesticide particles or other tiny, purpose-built structures having practical pesticide properties. A product is considered a nanopesticide if it uses nanoscale techniques to increase effectiveness or performance, such as employing components with at least one dimension between 1 and 100 nanometers, and is used to protect crops from a wide range of pests. These products, whose research is at various stages, can be used to increase the effectiveness of the active ingredients in pesticides now in use, raise their pollution control qualities, or even do both. In terms of pesticide use, nanopesticides are a relatively new technological development that offers a number of benefits, such as long lasting effectiveness and low amounts of potentially harmful active components. Many different formulation formats have been proposed, such as emulsions (such as nanoemulsions), nanocapsules (such as those made of polymers), and products produced with carefully crafted nanoparticles, such as metals, metal oxides, and nanoclays (Zaheer et al., 2022). The objectives of nanopesticide formulations are essentially the same as those of other pesticide formulations, namely (a) to raise the apparent solubility of poorly soluble AI or (b) to release the AI slowly and specifically and/or to prevent the AI from premature degradation. With the intention of studying potential effects on environmental fate, nanopesticides were initially categorized according to the intended use (Kah et al., 2013). In order to be effective, nanopesticides must either include extremely small quantities of the pesticide's active ingredient (AI) or other tiny designed structures. For the purposes of this study, a nanopesticide is a plant protection product that uses nanotechnology (such as the use of materials with a physical form that has at least one size dimension in the range of 1–100 nm) to improve the efficacy or lessen the environmental impact of a pesticide AI. It should be noted that the so-called "nano" formulations of proprietary insecticides frequently contain size fractions bigger than the conventional "nanoorange" of 100 nm. These products, which are in various stages of product development, can be used to boost the potency of present pesticide active ingredients, enhance their environmental sustainability safety profiles, or even do both (Kookana et al., 2014). In recent years, there has been an upsurge in interest in using nanotechnology in the development of plant protection products. The methodologies used to investigate the behavior and fate of nanopesticides in soils and water are currently the same as those used to understand the behavior and fate of conventional pesticide formulations. The standard approaches must be modified, however, in order to account for the unique characteristics of nanoformulations when assessing the environmental risk of nanopesticides (Lucia and Renato, 2020). A nanopesticide often consists of a pesticide active ingredient and a nanocarrier. The characteristics of both, as well as their interactions and linkages, have a significant influence on how soils and water behave and turn out. For instance, established methodologies accurately derive all researched processes by measuring the mass of the active component per volume of liquid or per mass of solid (AI concentrations). Other important considerations for pesticide nanoformulations include, but are not limited to, the interaction between the free AI and the AI attached to the nanocarrier (encapsulation efficiency), particle number concentration, particle size distribution, and others (Kookana et al., 2014). Building the best methods to comprehend and predict the fate of nanopesticides in soil and water requires an understanding of the function played by nanoparticles in the nanoformulation of pesticides and the predicted behavior that results from this involvement (Lucia and Renato, 2020).

2.2 DURABILITY OF ACTIVE INGREDIENTS OF NANOPESTICIDES

Knowing the longevity of the nanocarrier-AI complex after application in the environment is important in the case of controlled release of nanopesticides in order to set up the procedures for evaluating the fate of the AI in soils. There are three potential circumstances:

- Rapid AI release from the nanocarrier material (short durability). If the full release of the AI from the nanocarrier happens much faster than the relevant environmental process (degradation, leaching, runoff, and so forth) the exposure of the AI is comparable to that of formulations of conventional pesticides. As a result, there is little impact of the nanoformulation on AI behavior, and nanopesticides have the same environmental fate as pure AI.
- The nanocarrier material intermediate release of the AI. To evaluate the transfer of AI from the nanocarrier to environment, one must know the release kinetics of the nanopesticide. The rate of release, the characteristics of the AI in the nanocarrier and released AI form will all affect the environmental processes that are being researched in this context (Kah and Hofmann, 2014). As a result, it is anticipated that the nanoformulations will have an impact on how AI behaves in soil, necessitating the need for a more intricate exposure evaluation (Lucia and Renato, 2020).
- A negligible or slow release of the AI from the nanocarrier substance (long durability). The released AI's characteristics and the nanocarrier's colloidal characteristics alone will determine how the nanopesticides behave.

2.3 RELEASE EXPERIMENTS OF NANOPESTICIDES IN WATER

Two sorts of techniques can be employed to compare continuous strategies (membrane isolation techniques) and discontinuous ways for measuring the rates of release of nanopesticides in water (sampling-and-separate methods).

2.3.1 CONTINUOUS METHODS: DIALYSIS METHODS

In dialysis procedures, a semipermeable barrier physically separates the compartment containing the releasing medium (acceptor compartment) from the compartment containing AI in their nanoparticles (donor compartment). Pesticides that are not connected to the nanoparticles can pass through the membrane and reach the acceptor compartment, but the pesticide that is coupled to the nanoparticles cannot. The release kinetics are quite high for this approach compared to the diffusion rate via the membrane and inside the acceptor compartment, hence it is assumed that these components do not interfere with the release process. Adding the nanoformulation into a bag made of cellulose semipermeable membrane is the dialysis technique that is most frequently employed to ascertain the release kinetics of nanopesticides. The bag is then completely sealed before being submerged in a vessel that contains the releasing medium under sink conditions. The solubility of AI or the stability of the nanoformulation in various media and at various pH levels are a few parameters that are used to choose the release medium (Kumar et al., 2017). The amount of release medium needed to both maintain the sink conditions and propel the AI to the outside must be six to ten times larger than the volume of the dialysis bag. Since the unstirred water layer effect is reduced by agitation, pesticide diffusion from the dialysis bag to the vessel might be favored (D'Souza and DeLuca, 2006). In order to calculate the overall amount of pesticide released over time, samples from release media are periodically taken and tested. To maintain a constant total solution volume, an equivalent volume of new dissolving medium is added after each sampling in place of the release medium (Cao et al., 2018). An alternate approach uses a two-compartment model, where a glass vessel containing the nanopesticide is placed in a chamber that also contains the release medium while being covered by a dialyzing membrane (Silva et al., 2011). The dialysis approach has also been used to study the release (dissolution) of metal ions from metal nanoparticles. Copper chloride, which is a suitable chemical species for producing 65CuO NPs or nonisotopically modified CuO NPs, is available in isotopically enriched form (65Cu) (Misra et al., 2012).

2.3.2 DISCONTINUOUS METHODS: SAMPLE-AND-SEPARATE METHODS

A vessel holding the release medium is used in this procedure, and the nanopesticide is introduced before the release is monitored over time. The nanocarrier-AI complex is in direct touch with the bulk media, unlike dialysis procedures, samples, and distinct methods. The media is primarily chosen based on the AI's solubility, and the pH levels of this medium are chosen to assess the stability of the nanoformulation at various pH (Qian et al., 2011). Without sacrificing the sensitivity of the assay investigations, there should be enough release medium to sustain sink conditions (D'Souza and DeLuca, 2006). The system must be subjected to intermittent or constant agitation during the experiment after the nanoparticles have been incorporated into the release medium. On a regular basis, samples of the bulk medium are obtained, and the nanoparticles are filtered or centrifuged, separated, and then examined (Qian et al., 2013). To maintain sink conditions throughout the entire experiment, the release medium withdrawn must be replaced after each sampling with an equal volume of fresh release medium (Flores-Céspedes et al., 2012). An alternative technique setup was used, in which aliquots of the nanoformulation suspension sample were put to Falcon tubes that were sealed and stirred at room temperature after being added to the release medium (water) in the nanoformulation suspension sample. One of the tubes was taken out of the shaker at predetermined intervals and centrifuged. The supernatant was filtered and the pesticide release at each time was measured (Campos et al., 2015). This procedure is also used to assess the dilution rates (release rate) of metal ions from nanoparticles of the corresponding metal such as, for example, Ag or ZnO nanoparticles (Innovation and Island, 2010; Miller et al., 2010).

2.4 RELEASE EXPERIMENTS OF NANOPESTICIDES IN SOIL

In order to determine how long nanopesticides will last after being applied to a target area, it is essential to measure the release rate of AI from the nanocarrier in soils. The difficulties of collecting measurements at realistic soil-solution ratios, however, made the design of this experiment challenging (Kah et al., 2018). Similar methods for assessing the release of AI in soils have been put out in a number of works (Vickers, 2017). In beakers containing soil samples (25–50 g), the nanopesticide under test is wrapped in parchment paper and placed inside. The soil is then given more water until it can store about 60% more water. In a biochemical oxygen demand incubator set at 30°C, the beakers are covered with parafilm that has holes punched in it to facilitate air exchange. The soil-covered parchment paper strips are periodically removed, and beakers are periodically withdrawn from the incubator. The soil samples are extracted using the proper method and organic solvents (based on the AI), and the AI is then examined.

2.4.1 INDIRECT ESTIMATION

Because release tests may produce artefacts like it is occasionally impossible to directly assess durability due to the sorption of AI to membranes. In these cases, it may be able to infer the lifespan of nanopesticides indirectly by looking at other soil processes like persistence or leaching. An indirect technique for gauging the endurance of various nanoformulations of bifenthrin-loaded to polymer nanocarriers was established based on the kinetics of how rapidly free and connected (to the nanocarriers) AI deteriorate. The essential assumptions of this technique are that (1) only the released portion of the insecticide is available for degradation, (2) all insecticide was attached to the nanocarrier at time zero, and (3) the release and degradation processes are first order kinetically explainable. The authors suggested that the concentrations of formulated (nanocarrier-AI), released (AI), and degraded (AI') bifenthrin could be described by a sequential first-order model (Kah et al., 2016):

$$\text{Nanocarrier-AI} \xrightarrow{k1} \text{AI} \xrightarrow{k2} \text{AI}'$$

In this model, nanocarrier-AI represents the nanoformulation, AI refers to the active ingredient released from the nanocarrier, and AI' is the degraded active ingredient. The release rate of AI from nanocarriers is called $k1$; and the release rate of the active ingredient is called $k2$. Next, the findings from the degradation experiment with nanoformulations and pure AI were fitted to a sequential first-order model to determine the values of $k1$ and $k2$. The degradation experiment of the nanoformulations served as the basis for the calculation of k1, while the degradation curve of pure bifenthrin served as the basis for the calculation of $k2$. Finally, the release-half-lives (R50), or the amount of time needed for half of the bifentin to be released from the nanocarrier, were calculated from $k1$ (Table 2.1).

TABLE 2.1 Methodology Applied to Nanopesticides that Determines the Fate on NPs

Sl. No.	Nanopesticides	Methodology	References
1.	Chitosan loaded with hexaconazole	Release in water: strip procedure using parchment paper Release into the ground using strips of parchment paper	(Cabrera et al., 2016)
2.	2,4-Dichlorophenoxyacetic acid-loaded rice husk biochar	Sample-and-separate approach for release in water Release into the ground using strips of parchment paper	(Evy Alice Abigail, 2019)
3.	Poly(lactic acid) with abamectin added	Dialysis bag method for release in water	(Yu et al., 2017)
4.	Poly(ethylene glycol)/aliphatic and aromatic diacids loaded with thiamethoxam	Release into the ground using strips of parchment paper	(Taylor et al., n.d.)
5.	Alginate/chitosan loaded with acetamiprid	Release in water: strip procedure using parchment paper Release into the ground using strips of parchment paper	(Silva et al., 2011)
6.	Citrate-stabilized silver nanoparticles	Sample-and-separate approach for release in water	(Innovation and Island, 2010)
7.	Carboxymethyl-β-cyclodextrin-Fe_3O_4 magnetic nanoparticle loaded with diuron	Sample-and-separate approach for release in water	(Liu et al., 2014)
8.	Atrazine-loaded Poly(epsilon-Caprolactone)	Centrifugation and batch processes for sorption Release: indirect approach; Persistence: soil incubation experiment	(Pereira et al., 2014)
9.	Bifenthrin nanoformulations based on three different polymers	Batch technique for sorption Release in water: dialysis bag method; Persistence: soil incubation experiment	(Kah et al., 2016)
10.	Incorporated imidacloprid into poly(citric acid), poly(ethylene glycol), and poly(citric acid).	Release in water: dialysis bag method	(Memarizadeh et al., 2014)

2.5 TOXICOLOGICAL IMPACT OF NPS ON SOIL AND WATER

The size, concentration, and plant species of the nanoparticles in the pesticides at different stages of crop development all affect how toxic NPs are. Nanoparticles of nanopesticides' toxic effects are also influenced by their composition and size. Pesticides are mostly deposited in soil and sediments because they tend to settle there, which has an impact on how widely they are dispersed throughout the environment (Vryzas, 2018). Pesticides' adsorption into soil regulates both their mobility and permanence. Nanopesticides' enhanced solubility and bioavailability could have an impact on how they behave once they are absorbed by organisms, as well as how they behave in the environment (Iavicoli et al., 2017). The active ingredient's sorption to soils will be reduced, the compound's apparent persistence will be increased, and the active ingredient's uptake by earthworms and subsequent distribution will be changed (Anuar et al., n.d.). Fertility and soil biota may suffer if nanoparticles are utilized as fertilizers. They have an impact on soil bacteria, earthworm digestion, and microorganisms. Due to soil contamination brought on by nanoparticles, the properties of nanoparticles may alter the structure of the soil. Pesticide nanoparticles are easily released into the air or water and taken up by living things, causing hazardous effects on people, animals, and aquatic life. It worsens mortality, accelerates hatching, slows heartbeat, and interferes with proper growth. As a result, the composition, size (20 nm), and concentration (>100 ppm) of the nanoparticle are the major factors that determine the level of nanotoxicity in soil, plants, and water (Dubey and Mailapalli, 2016). Compared to other possible sources of nanoparticles like sewage sludge, nanopesticides have the potential to boost the flux of nanoparticles to soil by about three times. Due to their biomimetic characteristics and great capacity for diffusion and bioaccumulation in soil, aquatic habitats, and foods, nanopesticides have the potential to have toxicological impacts on all animals, particularly mammals (Nehra et al., 2020).

2.6 CONCLUSION

The usage of nanopesticides has drawn more attention recently. The methods used to evaluate how these novel nanoformulations behave in soils and water are almost identical to those employed for traditional pesticide formulations. The behavior of nanopesticides in soils and water, however, differs from that of conventional pesticides due to a variety of physicochemical characteristics related to their nano size and colloidal nature. Because of this, these processes have gradually been modified to accommodate nanoformulations. However, these modified methodologies have certain drawbacks when it comes to understanding the fate of nanopesticides. In spite of this, it is anticipated that new procedures will be developed that get around the drawbacks of traditional methods because of the increased interest in using nanotechnology in the formulation of plant protection products.

KEYWORDS

- **active ingredients**
- **degradation**
- **nanoformulations**
- **nanopesticides**
- **sewage**
- **sludge**
- **sorption**

REFERENCES

Anuar, M., Firdaus, M., Agatz, A., Hodson, M. E., Al-Khazrajy, O. S. A., & Boxall, A. B. A. (n.d.). Accepted Preprint. https://doi.org/10.1002/etc.4094.

Cabrera, A., Celis, R., & Hermosín, M. C. (2016). Imazamox-clay complexes with chitosan- and iron(III)-modified smectites and their use in nanoformulations. *Pest Management Science, 72*(7), 1285–1294. https://doi.org/10.1002/ps.4106.

Campos, E. V. R., De Oliveira, J. L., Da Silva, C. M. G., Pascoli, M., Pasquoto, T., Lima, R., Abhilash, P. C., & Fernandes Fraceto, L. (2015). Polymeric and solid lipid nanoparticles for sustained release of carbendazim and tebuconazole in agricultural applications. *Scientific Reports, 5*, 13809. https://doi.org/10.1038/srep13809.

Cao, L., Zhou, Z., Niu, S., Cao, C., Li, X., Shan, Y., & Huang, Q. (2018). Positive-charge functionalized mesoporous silica nanoparticles as nanocarriers for controlled 2,4-dichlorophenoxy acetic acid sodium salt release. *Journal of Agricultural and Food Chemistry, 66*(26), 6594–6603. https://doi.org/10.1021/acs.jafc.7b01957.

D'Souza, S. S., & DeLuca, P. P. (2006). Methods to assess *in vitro* drug release from injectable polymeric particulate systems. *Pharmaceutical Research, 23*(3), 460–474. https://doi.org/10.1007/s11095-005-9397-8.

Dubey, A., & Mailapalli, D. R. (2016). Nanofertilizers, nanopesticides, nanosensors of pest and nanotoxicity in agriculture. In *Nanotechnology in Agriculture* (pp. 307–330). https://doi.org/10.1007/978-3-319-26777-7.

Evy Alice Abigail, M. (2019). Biochar-based nanocarriers: Fabrication, characterization, and application as 2,4-dichlorophenoxyacetic acid nanoformulation for sustained release. *3 Biotech, 9*(8), 1–9. https://doi.org/10.1007/s13205-019-1829-y.

Flores-Céspedes, F., Figueredo-Flores, C. I., Daza-Fernández, I., Vidal-Peña, F., Villafranca-Sánchez, M., & Fernández-Pérez, M. (2012). Preparation and characterization of imidacloprid lignin-polyethylene glycol matrices coated with ethylcellulose. *Journal of Agricultural and Food Chemistry, 60*(4), 1042–1051. https://doi.org/10.1021/jf2037483.

Iavicoli, I., Leso, V., Beezhold, D. H., & Shvedova, A. A. (2017). Nanotechnology in agriculture: Opportunities, toxicological implications, and occupational risks. *Toxicology and Applied Pharmacology, 329*, 96–111. https://doi.org/10.1016/j.taap.2017.05.025.

Innovation, N., & Island, R. (2010). Ion release kinetics and particle persistence in aqueous nano-silver colloids. *Journal of Nanomaterials*, 2169–2175.

Kah, M., & Hofmann, T. (2014). Nanopesticide research: Current trends and future priorities. *Environment International, 63*, 224–235. https://doi.org/10.1016/j.envint.2013.11.015.

Kah, M., Beulke, S., Tiede, K., & Hofmann, T. (2013). Nanopesticides: State of knowledge, environmental fate, and exposure modeling. *Critical Reviews in Environmental Science and Technology, 43*(16), 1823–1867. https://doi.org/10.1080/10643389.2012.671750.

Kah, M., Kookana, R. S., Gogos, A., & Bucheli, T. D. (2018). A critical evaluation of nanopesticides and nanofertilizers against their conventional analogues. *Nature Nanotechnology, 13*(8), 677–684. https://doi.org/10.1038/s41565-018-0131-1.

Kah, M., Weniger, A. K., & Hofmann, T. (2016). Impacts of (nano)formulations on the fate of an insecticide in soil and consequences for environmental exposure assessment. *Environmental Science and Technology, 50*(20), 10960–10967. https://doi.org/10.1021/acs.est.6b02477.

Kookana, R. S., Boxall, A. B. A., Reeves, P. T., Ashauer, R., Beulke, S., Chaudhry, Q., Cornelis, G., Fernandes, T. F., Gan, J., Kah, M., Lynch, I., Ranville, J., Sinclair, C., Spurgeon, D., Tiede, K., & Van Den Brink, P. J. (2014). Nanopesticides: Guiding principles for regulatory evaluation of environmental risks. *Journal of Agricultural and Food Chemistry, 62*(19), 4227–4240. https://doi.org/10.1021/jf500232f.

Kumar, S., Kumar, D., & Dilbaghi, N. (2017). Preparation, characterization, and bio-efficacy evaluation of controlled release carbendazim-loaded polymeric nanoparticles. *Environmental Science and Pollution Research, 24*(1), 926–937. https://doi.org/10.1007/s11356-016-7774-y.

Liu, W., Yao, J., & Cai, M. (2014). Synthesis of a novel nanopesticide and its potential toxic effect on soil microbial activity. *Journal of Nanoparticle Research, 16*(5), 2677. https://doi.org/10.1007/s11051-014-2677-7.

Lucia, V., & Renato, S. S. D. C. (2020). Nanopesticides. In *Nanopesticides*. https://doi.org/10.1007/978-3-030-44873-8.

Memarizadeh, N., Ghadamyari, M., Adeli, M., & Talebi, K. (2014). Preparation, characterization and efficiency of nanoencapsulated imidacloprid under laboratory conditions. *Ecotoxicology and Environmental Safety, 107*, 77–83. https://doi.org/10.1016/j.ecoenv.2014.05.009.

Miller, R. J., Lenihan, H. S., Muller, E. B., Tseng, N., Hanna, S. K., & Keller, A. A. (2010). Impacts of metal oxide nanoparticles on marine phytoplankton. *Environmental Science and Technology, 44*(19), 7329–7334.

Misra, S. K., Dybowska, A., Berhanu, D., Croteau, M. N., Luoma, S. N., Boccaccini, A. R., & Valsami-Jones, E. (2012). Isotopically modified nanoparticles for enhanced detection in bioaccumulation studies. *Environmental Science and Technology, 46*(2), 1216–1222. https://doi.org/10.1021/es2039757.

Nehra, M., Dilbaghi, N., Marrazza, G., Kaushik, A., Sonne, C., Kim, K., & Kumar, S. (2020). Emerging nanobiotechnology in agriculture for the management of pesticide residues. *Journal of Hazardous Materials, 401*, 123369. https://doi.org/10.1016/j.jhazmat.2020.123369.

Pereira, A. E. S., Grillo, R., Mello, N. F. S., Rosa, A. H., & Fraceto, L. F. (2014). Application of poly(epsilon-caprolactone) nanoparticles containing atrazine herbicide as an alternative technique to control weeds and reduce damage to the environment. *Journal of Hazardous Materials, 268*, 207–215. https://doi.org/10.1016/j.jhazmat.2014.01.025.

Qian, K., Shi, T., He, S., Luo, L., Liu, X., & Cao, Y. (2013). Release kinetics of tebuconazole from porous hollow silica nanospheres prepared by miniemulsion method. *Microporous and Mesoporous Materials, 169*, 1–6. https://doi.org/10.1016/j.micromeso.2012.10.017.

Qian, K., Shi, T., Tang, T., Zhang, S., Liu, X., & Cao, Y. (2011). Preparation and characterization of nano-sized calcium carbonate as controlled release pesticide carrier for validamycin against *Rhizoctonia solani*. *Microchimica Acta, 173*(1, 2), 51–57. https://doi.org/10.1007/s00604-010-0523-x.

Silva, M. dos S., Cocenza, D. S., Grillo, R., Melo, N. F. S. de, Tonello, P. S., Oliveira, L. C. de, Cassimiro, D. L., Rosa, A. H., & Fraceto, L. F. (2011). Paraquat-loaded alginate/chitosan nanoparticles: Preparation, characterization and soil sorption studies. *Journal of Hazardous Materials, 190*(1–3), 366–374. https://doi.org/10.1016/j.jhazmat.2011.03.057.

Taylor, P., Sarkar, D. J., Kumar, J., Shakil, N. A., & Walia, S. (2012). Release kinetics of controlled release formulations of thiamethoxam employing nano-ranged amphiphilic PEG and diacid based block polymers in soil. *Journal of Environmental Science and Health, Part A: Toxic/Hazardous Substances and Environmental Engineering, 47*(11), 1701–1712. https://doi.org/10.1080/10934529.2012.687294

Vickers, N. J. (2017). Animal communication: When I'm calling you, will you answer too? *Current Biology, 27*(14), R713–R715. https://doi.org/10.1016/j.cub.2017.05.064.

Vryzas, Z. (2018). Pesticide fate in soil-sediment-water environment in relation to contamination preventing actions. *Current Opinion in Environmental Science & Health, 4*, 5–9. https://doi.org/10.1016/j.coesh.2018.03.001.

Yu, M., Yao, J., Liang, J., Zeng, Z., Cui, B., Zhao, X., & Sun, C. (2017). Poly(lactic acid) nanoparticles with regulatable release properties. *RSC Advances, 7*, 11271–11280. https://doi.org/10.1039/C6RA27345A.

Zaheer, T., Ali, M. M., Abbas, R. Z., Atta, K., Amjad, I., Suleman, A., Khalid, Z., & Aqib, A. I. (2022). Insights into nanopesticides for ticks: The superbugs of livestock. *Oxidative Medicine and Cellular Longevity, 2022*. https://doi.org/10.1155/2022/7411481.

CHAPTER 3

Nanopesticide Communication to Soil Microorganisms

NEELAM POONAR, POONAM MEENA, and ARCHANA MEENA

Department of Botany, University of Rajasthan, Jaipur, Rajasthan, India

ABSTRACT

Soil is composed of living organisms and serves as a habitat for microbial communities, which form the foundation of trophic food webs that sustain life on land. Soil microorganisms include bacteria, actinomycetes, fungi, protozoa, and nematodes, each of which contributes differently (both positively and negatively) to soil and plant health. Various external factors significantly influence the microbial community within the soil. However, plants and microorganisms are often overlooked because they are not the primary targets of pest control strategies. The critical role that soil microorganisms play in the proper functioning of ecosystems remains largely underappreciated. Most nanopesticides are systemic, meaning they are designed to act from within the plant. However, when released into the environment, nanopesticides affect non-target species, including soil microbiota, microfauna, and plants, which are exposed to the bioactive compounds present in these formulations.

3.1 INTRODUCTION

Access to high-quality food that is able to satisfy people's requirements across all continents is essential in light of the expanding human population. In conjunction with the pressure of economic investment in agricultural commodities, we have also witnessed an increase in the use

Nanopesticides: A Perspective from Welfare to Risk Assessment.
Garima Awasthi, Payal Lodha, and Gajendra Pal Singh (Eds.)

DOI: 10.1201/9781003605874-3

of fertilizers, insecticides, pesticides, and other agrochemicals to improve agricultural productivity. These agrochemicals aim to protect crops from pests and increase crop yields. Nevertheless, farmers, food producers, financing agencies, and governments need to be committed to striking a balance between the benefits and risks of cyclical declines generated by the indiscriminate use of agrochemical technologies in environments, ecosystems, consumers, and the entirety of the productive sector that is directly or indirectly involved in the agricultural chain (Kah and Hofman, 2013).

There are a large number of international patents and licensed products that connect farming peculiar technology and agriculture all over the world. These products concentrate on the provision of various commodities such as food, fuel, and wood for humankind (Yadav et al., 2020).

The purpose of agrochemical technology is generally to safeguard crop areas against pests (pathogens, harmful insects, and parasitic weeds) that compromise production and productivity. This includes diagnosing the vitality of soil and plantations, as well as controlling livestock and fishery products. In spite of this, the indiscriminate use of pesticides against harmful insects and rodents has also had a negative impact on production, leading to resilience to pathogens and insects, driving up demand for new agrochemicals, and worsening environmental imbalance (Bombo et al., 2019).

Because of the increased total surface area that agrochemical nanoparticulates occupy, there is an overall greater interaction with crop pests. This makes nanoparticulates equally effective and unpredictable. Because of the unreasonable and unsystematic use of agrichemicals, pathogen resistance is increased, nitrogen fixation and biodiversity are decreased, and the bioaccumulation of pesticides in agricultural goods, livestock goods, and organisms of the water environment is increased. This poses a severe and growing threat not only to the ecosystem but also to humans and is a barrier to the development of sustainable agriculture (Osorio-Echavarria et al., 2021; Vigneshwaran et al., 2006).

In this chapter, the effects of nanopesticides on soil microbes have been discussed.

3.2 SOIL MICROBES

Soil is made of living things and is where microbial communities live. Microbial communities are the basis of trophic food webs, which support life on land. Up to 1012 bacteria and 2500 of fungi can be found in a healthy soil. But because cells tend to stick together, soil microbes only cover about 10–6% of the soil's surface area (Young et al., 2008).

Bacteria, actinomycetes, fungus, protozoa, and nematodes are soil microorganisms. Each microorganism type boosts soil and plant health differently.

3.2.1 BACTERIA

Soil bacteria are essential. They break down nutrients and release them to the plant's roots. The FAO stated bacteria may be the most valuable soil life form.

3.2.2 ACTINOMYCETES

Actinomycetes were originally considered to be a type of fungi due to their comparable behavior in soil. However, certain actinomycetes are harmful to the plant because they are predators, while other actinomycetes that live in the soil have the ability to treat the plant with antibiotics.

3.2.3 FUNGI

Fungi, like bacteria, can be found in the root zone of plants and are crucial in the process of making nutrients accessible to plant roots. For instance, mycorrhizae is a type of fungi that helps plant roots absorb water and nutrients, which in turn supplies the plant with carbohydrates, amino acids, and other essential elements.

3.2.4 PROTOZOA

Protozoa are larger germs that adore consuming bacteria and living in environments that are filled with them. In point of fact, the release of nutrients that were previously used by bacteria occurs when protozoa consume the bacteria.

3.2.5 NEMATODES

Nematodes are very small worms that can be found either outside of or inside of plants. There are nematodes that prey on other nematodes, and there are nematodes that are useful to plants because they consume pathogenic nematodes and secrete nutrients.

3.3 BENEFITS OF SOIL MICROBES

In soil, many microorganisms are pathogenic which impact crop production negatively while many soil microbes have various benefits such as:

1. These can degrade organic matter.
2. They help in the recycling of nutrients by the process of demineralization.
3. They can produce oxygen gas.
4. They help in nitrogen fixation.
5. By the process of immobilization, they help in the maintenance of nutrients in cells.
6. By interacting with other organisms, they release nutrients.
7. They help in the absorption of nutrients by plant roots.
8. They help in soil structure and water holding.
9. They help in the degradation of soil pollutants.

3.4 FACTORS AFFECTING SOIL MICROBIAL COMMUNITY

The major external factors that influence the microbial community in soil are discussed below.

3.4.1 SOIL MOISTURE

The porosity of the soil determines the amount of water that may be held within it. The size of the pores in the soil is determined by its texture, which is defined as the ratio of sand, silt, and clay particles. In addition, soil moisture can be altered by management practices like crop rotation and tillage, as well as by irrigation and drainage systems, all of which serve to improve the soil's capacity to absorb and transmit water.

3.4.2 SOIL ORGANIC MATTER

The decomposing organic matter of plant and animal origin that has died serves as the total soil organic matter, which is then colonized by microbes and broken down by them. The size of the microbial population in the soil, on the other hand, grows as green manures, agricultural leftovers, and other organic matter are worked into the ground.

3.4.3 TYPES OF VEGETATION AND ITS GROWTH STAGES

It is possible that one of the two groups will predominate depending on the type of vegetation present and the stage the plant is in its life cycle, according to the season and growth stages, respectively.

3.4.4 DIFFERENT SEASONS

The quantity of microorganisms present and their level of activity are the primary factors that determine the amount of nutrients that are available to plants. They continue to exist in a condition of continuous and unpredictability in the soil, which is a medium in which the microbial community is subject to a significant deal of physicochemical and biological influence. It is well established that there are shifts in the microbial communities found in the soils of tropical, subtropical, and temperate climates.

3.4.5 ARCHITECTURE OF SOIL

There is a clear relationship between the microstructure of soil aggregates and the biodiversity of soil ecosystems and functions. The diameter of the pore spaces can control the diffusion of gas, water, and nutrients. This diameter ranges from 10 to 30 micrometers. Microorganisms can take advantage of the microniches provided by the soil's microstructure. For instance, the nitrogen cycle is dependent on different communities of organisms inhabiting different parts of the structure of soil. Nitrifier is more prevalent and activated in microaggregates ranging in size from 2 to 20 micrometers, but N-fixing microbes were found to predominate in the soil. All of these architectural structures in the soil are responsible for controlling the interactions that take place between plants, microorganisms, and pollutants. For this reason, it will be extremely crucial to gain a knowledge of the interactions that occur between the soil matrix and nanopesticides.

3.4.6 ORGANIC AND INORGANIC CHEMICALS

Because these compounds supply sustenance for the development, activity, and survival of microbes in ecologically inadequate niches in soil, they are particularly important for microorganisms. The chemical components include things like gases, acids, trace and major elements, clay minerals, and so on. There are microorganisms and various gases, including oxygen, methane, and carbon dioxide, that are dissolved in the soil solution.

There is an intricate feedback loop that exists between the chemical properties of the matters and the biological properties of the microbes that make their home in the soil. On a smaller scale, the soil can be thought of as a three-dimensional hierarchical network that is made up of aggregates and pores. These networks are intermittently connected by wetting events. The aggregate is the fundamental component of the ecosystem that makes up the soil (Wilpiszeski et al., 2019).

Association with organic matter and minerals cause the formation of cluster that are between 2 and 20 microns in size, which act as hutches for bacteria and fungi (Watteau and Villemin, 2018). Plant exudates and desquamated cells increase hot spots of bacteria, which in turn encourage the establishment of persistent clusters at the rhizosphere, which is the contact between the plant root and the soil (Watteau and Villemin, 2018). The pores and microaggregates (250 mm in size) are formed when these clusters coalesce. This architecture produces a varied flow of water and nutrients that the organisms living in the soil are able to access (Wilpiszeski et al., 2019). Enzymatically active proteins can be found anywhere, including inside of cells, inside or on the surface of microaggregates and macroaggregates, and even in pores when there is waterlogging present. The capacity of a soil to interact with microorganisms is referred to as its "cation exchange capacity." Soil, particularly notably clays, organic matter, and minerals, have the ability to sorb chemical compounds that are circulating in the pore water solution.

In this chapter, we will discuss about communication of nanopesticides (under the category of organic/inorganic chemicals) on soil microbes.

3.5 EFFECTS OF NANOPESTICIDES ON SOIL MICROORGANISMS (NON-TARGET)

Nanopesticides are described as preparations that purposely introduce components in the nm size range. Since the properties related with this small size range are new, these nanopesticides have been introduced to the market in the past. Nanopesticides can't be lumped into a single category because they've already been used into so many different goods.

Plants and microorganisms are not given the same level of consideration because they are not the intended targets. The contribution that soil microorganisms provide to the proper functioning of ecosystems is still grossly underappreciated.

The term "nanopesticide" can refer to a variety of different forms of pesticides, including nanoformulations of traditional pesticides as well as nanomaterials that are used as pesticides; the majority of these nanoparticles are metallic or metal oxide.

3.5.1 EFFECT OF NANOPESTICIDES BASED ON METAL AND METAL OXIDE NANOPARTICLES

The vast majority of nanopesticides are systemic, meaning that they are designed to work from within the plant. However, as nanopesticides are released into the environment, the microbiota and microfauna of the soil, as well as the plants themselves, become non-target species that are subject to the effects of the bioactive compounds.

The effects of metal and metal oxide nanoparticles on the microbiome of soil and rhizosphere have been the subject of a significant amount of research, mostly for the purpose of determining how they might contribute to environmental contamination (Anjum et al., 2013; Simonin and Richaume, 2015; Rajput et al., 2020). Nanoparticles derived from TiO_2, Ag, ZnO, Cu, and Fe dominate the list of substances that have been subjected to the most research in toxicity studies.

3.5.2 EFFECTS OF COPPER COATED NANOPESTICIDES ON SOIL MICROBES

Copper micronized particles were initially marketed, but copper nano-sized particles have been rapidly created and commercialized to improve coverage of plant fruits or leaves and to control the release of copper ions. At this time, there are at least two nanosized copper formulations available on the market: Kocide® 3000 (manufactured by DuPont) and NANOCU (manufactured by Bio Nano Technology) (He et al., 2019).

Various studies have been revealed the effects of nanopesticides on soil microbes which have been summarized in Table 3.1.

TABLE 3.1 Effect of Copper Nanopesticides on Soil Microbes

Sl. No.	Nanopesticide	Observed Effect	Researchers
1.	Bare Copper oxide (50 nm,) vs ($CuSO_4$)	CuONPs caused significant reductions that increased with time, on soil microbial activities involved in carbon and nitrogen cycles, respiration, nitrification, and denitrification.	Simonin et al. (2018a)
2.	Cu-based NPs	Cu-based NPs showed toxicity to nitrifiers and the very different behaviour between CuONPs and Cu^{2+} ions in terms of Cu^{2+} release, adsorption and impact on nitrification in batch nitrification kinetic experiments.	VandeVoort and Arai (2012)
3.	Nanosized CuO and Cu ions	The activities of 5 soil extracellular microbial enzymes involved in C, N and P nutrient cycling decreased at 2 h which was recovered after 30 days of application.	Asadishad et al. (2018)
4.	Kocide® 3000 (Dupont) is fungicide/bactericide based on copper hydroxide	Act as fungicide/bactericide based on copper hydroxide, for citrus, conifers, field crops, small fruits, tree crops, vegetables, vines and some other fruits.	Adeleye et al. (2014)
5.	Kocide® 3000 (particle size was 38.7 ± 8.2 nm)	dual, beneficial or negative, interactive effects between nanopesticide and fertilization treatments on extracellular microbial enzymatic activities.	Simonin et al. (2018b)
6.	Commercial $Cu(OH)_2$ nanopesticide formulation	$Cu(OH)_2$ nanopesticides at 50 $mg.kg^{-1}$ significantly decreased the alpha-diversity of bacteria in soil and drastically altered the community composition	Zhang et al. (2019)
7.	CuNPs (40–60 nm)	bacteria, fungi and nitrifying bacteria, remained relatively stable throughout the experiment (at 0.5–0.15%) However, CuNPs at 0.15% w/w caused a significant decrease in soil.	Parada et al. (2019)

TABLE 3.1 *(Continued)*

Sl. No.	Nanopesticide	Observed Effect	Researchers
8.	CuONPs (hydrodynamic diameter in water 240.0 nm) and CuO bulk particles (BP, average particles size of 1346 nm)	The microbial biomass carbon in both soils was severely inhibited by CuO NPs and to a minor extent by BPs at 100 mg.kg^{-1}. The organic matter could partly mitigate the negative effects of CuO NPs.	Shi et al. (2018)

3.5.3 EFFECTS OF SILVER COATED NANOPESTICIDES ON SOIL MICROBES

Since ancient times, people have used silver as a biocide. According to Durán et al. (2016), silver-based nanopesticides exhibit antibacterial and biocidal effects against a wide variety of different types of microorganisms, including bacteria, fungus, and viruses.

There are already certain silver-based nanopesticides that have been patented and are commercially available. Some examples include WA-CV-WA13B, WA-AT-WB13R, and WA-PR-WB13R, all of which are manufactured by Bio-Plus Co., Ltd. in Pohang, Korea; in addition, there is Zerebra® agro, Zeroxxee®, Silver leaf, and Zeromix®.

Effect of silver coated nanopesticides on soil microbes have been studied by many researchers which are summarized in Table 3.2.

TABLE 3.2 Effects of Solver Nanopesticides on Soil Microbes

Sl. No.	Nanopesticide	Observed Effects	Researchers
1	AgNPs (NM-300K dispersed in a mixture of stabilizing agents, particle size ~15 nm, 99%)	Next-generation sequencing evidenced varying sensitivities among bacterial orders, with a shift in the microbial community. Enrichment of *Proteobacteria* (including *Caulobacterales*, *Burkholderiales*, and *Xanthomonadales*), *Cytophagales*, and *Sphingobacteriales* was observed. The adverse impact on some nitrifiers (*Nitrosomonadales*) corresponded with the inhibition of PAO activity.	Hund-Rinke et al. (2019)
2	AgNPs (NM-300K, diameter ~15 nm, and a small proportion at ~95 nm)	AgNPs showed relative inhibition of ammonia-oxidizing bacteria (AOB), starting from day 14, with increasing inhibition up to 140 days.	Schlich et al. (2018)

TABLE 3.2 *(Continued)*

Sl. No.	Nanopesticide	Observed Effects	Researchers
3	Zerebra® Agro Ag nanoparticles (1 mg/L, average particle size ~25 nm, specific surface area 23 m^2/g, dispersed in water with a soap surfactant)	AgNPs treatment reduced aerobic bacteria and stimulated facultative anaerobic and oxidative stress-tolerant bacteria.	Vitali et al. (2019)
4	AgNPs (50 nm citrate-coated AgNPs)	The relative abundance of the *Gammaproteobacteria* class was significantly higher for Ag+ ions and AgNPs at 100 mg/kg soil compared to other treatments. The *Alphaproteobacteria* community showed a decrease in relative abundance with dissolved Ag at 100 mg/kg soil.	Asadishad et al. (2018)
5	AgNPs (BAM-N001 AgPure)	The negative effects were most pronounced for *beta-Proteobacteria* and *Bacteroidetes*. *Actinobacteria* and *alpha-Proteobacteria* were statistically unaffected by AgNP treatments after 1-year exposure.	Grün et al. (2018, 2019)
6	AgNPs (spherical nanoparticles, size distribution between 20 and 30 nm by scanning electron microscopy, 0.15×10^{12} and 0.31×10^{12} NPs/mL)	The coating of the nanoparticles likely delayed the release of Ag+, explaining the latency phase in the changes in bacterial abundance and genes involved in the nitrogen cycle. Biogenic AgNPs tended to stimulate bacteria involved in the nitrogen cycle, showing alternating cycles of impact and recovery.	Guilger et al. (2017)
7	AgNPs (PVP-coated, 50 nm and 15 nm)	The smallest PVP-coated AgNPs (15 nm) showed the greatest differences from the control condition, with a concentration-dependent inhibition of the reaction rate.	VandeVoort et al. (2014)
8	Ag_2S NPs, polyvinylpyrrolidone (PVP)-coated AgNPs	The microbial community was affected even at 1 mg/kg for Ag-PVP NPs, Ag+, and Ag_2S NPs, with impacts observed on fungi and bacteria, including *Actinomycetes*.	Judy et al. (2015)

3.5.4 EFFECTS OF OTHERS NANOPESTICIDES BASED ON INORGANIC NANOMATERIALS

Other nanopesticides are envisaged, based on nanomaterials of TiO_2, ZnO, CeO_2, SiNPs and even carbon nanotubes. Many studied have also shown the environmental impacts on microbiota of these NPs (Liné et al., 2017 and Tian et al., 2019).

3.6 METHODS TO ANALYSE IMPACT OF NANOPESTICIDES ON SOIL MICROORGANISMS

3.6.1 ANALYSIS OF MICROBIOME

Analysis of the microbiome is dependent on metagenomics and, more generally, omics (transcriptomics, proteomics, and metabolomics), which removes the need to cultivate bacteria in order to study them in their natural setting. The composition, abundances, and activities of microbial communities, such as the RNA, proteins, and metabolites that they produce, can be used to characterize these communities.

Amplicon sequencing, also known as "DNA metabarcoding," is one method that can be used to define the microbiome (Caporaso et al., 2011; Shokralla et al., 2012).

Metagenomics is a method that can be used to delve more deeply into the study of microbiomes. This method examines the complete set of genes that are present, which not only leads to the composition of the microbiome, but also to the complete set of functions that the microbiome is capable of displaying.

3.6.2 EVALUATION OF MICROBIAL ENZYMATIC ACTIVITY

There is a potential for changed biochemical processes that are driven by microorganisms to result from the interaction of pesticides with soil. Enzymes can be found in the soil in many different forms, including free, immobilized, extracellular, and intracellular entities. According to Garcia-Ruiz et al. (2008) and Karaca et al. (2011), soil enzyme activity are indicators of soil quality. Soil enzymes play multiple functions in the cycling of nutritional elements and the decomposition of organic waste. Therefore, the activities of soil enzymes are useful biological responses for analyzing the response of the soil to a stress such as the presence of pesticides. In the process of element cycling, soil enzymes play an essential part in the C cycle, the N cycle, the P cycle (which involves phosphatases), and the S cycle (arylsulfatase). Intracellular enzymes known as dehydrogenases can be present in all organisms that are alive and play a role in the transfer of energy during microbial metabolic events as well as the biological oxidation of organic matter in soil. According to Wolska et al. (2015), they are frequently utilized as an indicator of the overall microbial activity in soil.

3.7 DISCUSSION

The use of nanotechnology holds the potential to reduce the harmful effects of pesticides on soil microorganisms that are not their intended target. There is a significant amount of untapped potential in altering the surface of the NPs in order to tune their characteristics, their interactions, and their eventual fate in the soil. Encapsulating active chemicals in polymers, formulating biogenic nanoparticles, and tailoring the characteristics of NPs to limit their impact are all potentially fruitful avenues of investigation.

Utilizing environment stimuli responsive nanopesticides is one way that smart nanoparticulate vectors of pesticides, which are currently in the process of being developed but have the potential to be used in the near future, could be designed to deal with controlled and targeted pesticide release. All of these intelligent devices ought to rely on eco-friendly technology and substances that are compatible with living organisms.

Controlling the spread of nanopesticides and determining whether or not they are harmless to the microorganisms that live in soil is an essential step in the process of developing nanopesticides. This is necessary in order to protect the ecology of the soil. For the purpose of preventing pollution of the water compartment, within the soil matrix. The use of synthetic virus-based model nanopesticides whose mobility permits them to reach varied depths in soil could be envisioned for the purpose of soil depth targeted release (Chariou et al., 2019).

Research is currently centered on the hunt for microbiota that will either enable plants to defend themselves against biotic (plant diseases) or abiotic (drought, flooding, and so forth) challenges, or boost the growth and output of field crops. Nanopesticides need to be able to fit into this framework and permit the combined use of both preventative (seed treatment, disinfection, or promotion of seedling transplanting) and curative (plant therapy) approaches to pest control.

In terms of the effect that it has on the microbiome, the methodology focuses on the diversity that is shown by sequencing an amplicon of 16S rRNA. This is in relation to the effect that it has on the bacterial population that is already there. A more comprehensive strategy would take into account the diversity of bacteria in addition to that of archaea, fungi, protozoa, and other organisms. This would make it possible to investigate how nanopesticides are dismantling the networks of relationships that exist between different communities. A more significant step forward may involve concentrating on active communities (complementary DNA) as well as the expression of genes. In point of fact, extracellular DNA can survive in soil for long periods of time and conceal some community shifts.

The effect that nanopesticides have on the microbiota of the soil can also be altered by some species that live in the soil, such as nematodes. Recent research conducted by Bart et al. (2019) provided evidence that nematodes have the ability to reduce the toxic effects of pesticides on the enzymatic activities of soil microbes.

Returning to soil, which serves as the fundamental matrix for agronomy, it should be noted that the microstructure of soil aggregates has a direct influence on the communities and functional diversity of soil, as well as the likely implications of nanoagrochemicals. Microfluidic techniques offer new approaches to the study of soil microbial ecology, making it possible to simulate and manipulate chemical conditions and physical structures at the microscale in soil model habitats. This helps researchers get around the challenges posed by the complexity of the soil matrix (Aleklett et al., 2018).

The accumulation of these findings suggests that copper and silver nanoparticles have the potential to radically alter the make-up of microbial communities and the activity of extracellular enzymes that are involved in the cycling of elements. The behavior and destiny of Cu- and AgNPs in soil is dependent on variables that are both intrinsic to the NPs themselves, such as particle size, surface charge, and isoelectric point (the pH at which the NPs do not carry a net electrical charge), as well as extrinsically related to the characteristics of the complex soil matrix. The dissolution of nanoparticles and their interactions with cells are both heavily influenced by the shapes of the nanoparticles themselves. Depending on the amount of AgNPs or NPs that an organism is exposed to, the properties of these nanoparticles can have a variety of effects on the microbial community's make-up as well as its functions (Zhai et al., 2016).

Denitrification is at the top of the list of microbial processes that NPs have the potential to influence. At the microscopic scale, the structural arrangement of the soil creates a variety of niches that are conducive to the growth of bacteria that have varied requirements (aerobic or anaerobic) and ways of life. These findings suggest that the microaggregate The absence of oxygen, the presence of nitrate, and the presence of carbon (a source of electrons) are the primary factors that have a favorable impact on the process of denitrification (Zumft, 1997). In addition to this, denitrifies are sensitive to pH (Imek and Cooper, 2002) as well as hydric environments (Szukics et al., 2010). Denitrification is more likely to occur in soils with lower redox potential values, which are connected to the soil's texture (Kunickis et al., 2010). Denitrification cannot occur in sandy-textured soils because their redox values are typically too high for this biological process. This demonstrated that the properties of AgNPs could be tuned to avoid impacts on denitrification. As a result, the biogenic AgNPs that were produced by an environmentally friendly approach (Guilger et al., 2017) did not demonstrate a significant impact on the nitrogen cycle.

3.8 CONCLUSION AND FUTURE DIRECTIONS

In conclusion, we might say that soil should be thought of as a kind of superorganism. The ecology of the soil needs to be considered on a global scale and in interaction with the air and water compartments before intelligent solutions can be designed for the field of agronomy.

Long-term studies have shown that the toxicity of NPs is kinetic, and it appears to be related to dissolution or transformation processes in the soil. These events lead to temporary adjustments and adaptations of the microbial population. Tuning the surface properties of NPs could help to control the dissolution and phase changes, and could also lower the toxicity against microbial cells.

A promising direction most likely relies on biogenic nanopesticides, which exhibit a limited impact on human cells and denitrification, but considerable activity toward a collection of plant pathogenic fungi.

KEYWORDS

- **bioactive compounds**
- **denitrification**
- **microbial communities microbiome**
- **nanopesticides**
- **soil microorganisms**

REFERENCES

Adeleye, A. S., Conway, J. R., Perez, T., Rutten, P., & Keller, A. A. (2014). Influence of extracellular polymeric substances on the long-term fate, dissolution, and speciation of copper-based nanoparticles. *Environmental Science & Technology, 48*(22), 12561–12568.

Aleklett, K., Kiers, E. T., Ohlsson, P., Shimizu, T. S., Caldas, V. E., & Hammer, E. C. (2018). Build your own soil: Exploring microfluidics to create microbial habitat structures. *The ISME Journal, 12*, 312–319.

Allen, H. K., Donato, J., Wang, H. H., Cloud-Hansen, K. A., Davies, J., & Handelsman, J. (2010). Call of the wild: Antibiotic resistance genes in natural environments. *Nature Reviews Microbiology, 8*, 251–259.

Anjum, N. A., Gill, S. S., Duarte, A. C., Pereira, E., & Ahmad, I. (2013). Silver nanoparticles in soil–plant systems. *Journal of Nanoparticle Research, 15*, 1896.

Asadishad, B., Chahal, S., Akbari, A., Cianciarelli, V., Azodi, M., Ghoshal, S., & Tufenkji, N. (2018). Amendment of agricultural soil with metal nanoparticles: Effects on soil enzyme activity and microbial community composition. *Environmental Science & Technology, 52*, 1908–1918.

Bart, S., Pelosi, C., Barraud, A., Péry, A. R. R., Cheviron, N., Grondin, V., Mougin, C., & Crouzet, O. (2019). Earthworms mitigate pesticide effects on soil microbial activities. *Frontiers in Microbiology, 10*. https://www.frontiersin.org/articles/10.3389/fmicb.2019.01535/full (accessed on 3 October 2024).

Bombo, A. B., Pereira, A. E. S., Lusa, M. G., De Medeiros-Oliveira, E., De Oliveira, J. L., Campos, E. V. R., De Jesus, M. B., Oliveira, H. C., Fraceto, L. F., & Mayer, J. L. S. (2019). A mechanistic view of interactions of a nanoherbicide with target organisms. *Journal of Agricultural and Food Chemistry, 67*, 4453–4462.

Camara, M. C., Campos, E. V. R., Monteiro, R. A., do Espirito Santo Pereira, A., de Freitas Proença, P. L., & Fraceto, L. F. (2019). Development of stimuli-responsive nano-based pesticides: Emerging opportunities for agriculture. *Journal of Nanobiotechnology, 17*, 100.

Caporaso, J. G., Lauber, C. L., Walters, W. A., Berg-Lyons, D., Lozupone, C. A., Turnbaugh, P. J., Fierer, N., & Knight, R. (2011). Global patterns of 16S rRNA diversity at a depth of millions of sequences per sample. *Proceedings of the National Academy of Sciences, 108*, 4516–4522.

Chariou, P. L., Dogan, A. B., Welsh, A. G., Saidel, G. M., Baskaran, H., & Steinmetz, N. F. (2019). Soil mobility of synthetic and virus-based model nanopesticides. *Nature Nanotechnology, 14*, 712–718.

Durán, N., Durán, M., de Jesus, M. B., Seabra, A. B., Fávaro, W. J., & Nakazato, G. (2016). Silver nanoparticles: A new view on mechanistic aspects of antimicrobial activity. *Nanomedicine: Nanotechnology, Biology and Medicine, 12*, 789–799.

García-Ruiz, R., Ochoa, V., Hinojosa, M. B., & Carreira, J. A. (2008). Suitability of enzyme activities for the monitoring of soil quality improvement in organic agricultural systems. *Soil Biology and Biochemistry, 40*, 2137–2145.

Grün, A. L., Manz, W., Kohl, Y. L., Meier, F., Straskraba, S., Jost, C., Drexel, R., & Emmerling, C. (2019). Impact of silver nanoparticles (AgNP) on soil microbial community depending on functionalization, concentration, exposure time, and soil texture. *Environmental Sciences Europe, 31*, 15.

Grün, A. L., Straskraba, S., Schulz, S., Schloter, M., & Emmerling, C. (2018). Long-term effects of environmentally relevant concentrations of silver nanoparticles on microbial biomass, enzyme activity, and functional genes involved in the nitrogen cycle of loamy soil. *Journal of Environmental Sciences, 69*, 12–22.

Guilger, M., Pasquoto-Stigliani, T., Bilesky-Jose, N., Grillo, R., Abhilash, P. C., Fraceto, L. F., & Lima, R. (2017). Biogenic silver nanoparticles based on *Trichoderma harzianum*: Synthesis, characterization, toxicity evaluation, and biological activity. *Scientific Reports, 7*, 44421.

He, X., Deng, H., & Hwang, H. (2019). The current application of nanotechnology in food and agriculture. *Journal of Food and Drug Analysis, 27*, 1–21.

Hund-Rinke, K., Hümmler, A., Schlinkert, R., Wege, F., & Broll, G. (2019). Evaluation of microbial shifts caused by a silver nanomaterial: Comparison of four test systems. *Environmental Sciences Europe, 31*, 86.

Kah, M., & Hofmann, T. (2014). Nanopesticide research: Current trends and future priorities. *Environmental International, 63*, 224–235.

Karaca, A., Cetin, S. C., Turgay, O. C., & Kizilkaya, R. (2011). Soil enzymes as indication of soil quality. In G. Shukla & A. Varma (Eds.), *Soil Enzymology* (pp. 119–148). Springer.

Kędziora, A., Speruda, M., Krzyżewska, E., Rybka, J., Łukowiak, A., & Bugla-Płoskońska, G. (2018). Similarities and differences between silver ions and silver in nanoforms as antibacterial agents. *International Journal of Molecular Sciences, 19*(3).

Kunickis, S. H., Gilliam, J. W., Evans, R. O., & Dukes, M. (2010). Soil characteristics and their role in developing conditions favorable for denitrification. In *Proceedings of the 19th World Congress of Soil Science: Soil Solutions for a Changing World* (pp. 34–37). Brisbane, Australia.

Lensi, R., Clays-Josserand, A., & Jocteur Monrozier, L. (1995). Denitrifiers and denitrifying activity in size fractions of a mollisol under permanent pasture and continuous cultivation. *Soil Biology and Biochemistry, 27*(1), 61–69.

Liné, C., Larue, C., & Flahaut, E. (2017). Carbon nanotubes: Impacts and behavior in the terrestrial ecosystem: A review. *Carbon, 123*, 767–785.

Marshall, B. M., & Levy, S. B. (2011). Food animals and antimicrobials: Impacts on human health. *Clinical Microbiology Reviews, 24*(4), 718–733.

Millardet, A., Gayon, U., & Schneiderhan, F. J. (1933). The discovery of Bordeaux mixture. Three papers: I. Treatment of mildew and rot. II. Treatment of mildew with copper sulphate and lime mixture. III. Concerning the history of the treatment of mildew with copper sulphate. *American Phytopathological Society*.

Osorio-Echavarría, J., Osorio-Echavarría, J., Ossa-Orozco, C. P., & Gómez-Vanegas, N. A. (2021). Synthesis of silver nanoparticles using white-rot fungus *Anamorphous Bjerkandera sp. R1*: Influence of silver nitrate concentration and fungus growth time. *Scientific Reports, 11*(1), 1–14.

Parada, J., Rubilar, O., Sousa, D. Z., Martínez, M., Fernández-Baldo, M. A., & Tortella, G. R. (2019). Short term changes in the abundance of nitrifying microorganisms in a soil-plant system simultaneously exposed to copper nanoparticles and atrazine. *Science of the Total Environment, 670*, 1068–1074.

Rajput, V., Minkina, T., Ahmed, B., Sushkova, S., Singh, R., Soldatov, M., & Musarrat, J. (2020). Interaction of copper-based nanoparticles to soil, terrestrial, and aquatic systems: Critical review of the state of the science and future perspectives. In P. de Voogt (Ed.), *Reviews of Environmental Contamination and Toxicology* (Vol. 252, pp. 51–96). Springer International Publishing.

Santaella, C., Schue, M., Berge, O., Heulin, T., & Achouak, W. (2008). The exopolysaccharide of *Rhizobium sp. YAS34* is not necessary for biofilm formation on *Arabidopsis thaliana* and *Brassica napus* roots but contributes to root colonization. *Environmental Microbiology, 10*(8), 2150–2163.

Schlich, K., Hoppe, M., Kraas, M., Schubert, J., Chanana, M., & Hund-Rinke, K. (2018). Long-term effects of three different silver sulfide nanomaterials, silver nitrate and bulk silver sulfide on soil microorganisms and plants. *Environmental Pollution, 242*, 1850–1859.

Seiler, C., & Berendonk, T. U. (2012). Heavy metal driven co-selection of antibiotic resistance in soil and water bodies impacted by agriculture and aquaculture. *Frontiers in Microbiology, 3*, 399.

Shi, J., Ye, J., Fang, H., Zhang, S., & Xu, C. (2018). Effects of copper oxide nanoparticles on paddy soil properties and components. *Nanomaterials, 8*(12), 977.

Shokralla, S., Spall, J. L., Gibson, J. F., & Hajibabaei, M. (2012). Next-generation sequencing technologies for environmental DNA research. *Molecular Ecology, 21*(8), 1794–1805.

Šimek, M., & Cooper, J. E. (2002). The influence of soil pH on denitrification: Progress towards the understanding of this interaction over the last 50 years. *European Journal of Soil Science, 53*(3), 345–354.

Simonin, M., & Richaume, A. (2015). Impact of engineered nanoparticles on the activity, abundance, and diversity of soil microbial communities: A review. *Environmental Science and Pollution Research, 22*(17), 13710–13723.

Simonin, M., Cantarel, A. A. M., Crouzet, A., Gervaix, J., Martins, J. M. F., & Richaume, A. (2018). Negative effects of copper oxide nanoparticles on carbon and nitrogen cycle microbial activities in contrasting agricultural soils and in presence of plants. *Frontiers in Microbiology, 9*, 1924.

Szukics, U., Abell, G. C. J., Hödl, V., Mitter, B., Sessitsch, A., & Zechmeister-Boltenstern, S. (2010). Nitrifiers and denitrifiers respond rapidly to changed moisture and increasing temperature in a pristine forest soil. *FEMS Microbiology Ecology, 72*(3), 395–406.

Tian, H., Kah, M., & Kariman, K. (2019). Are nanoparticles a threat to mycorrhizal and rhizobial symbioses? A critical review. *Frontiers in Microbiology, 10*, 1513.

VandeVoort, A. R., & Arai, Y. (2012). Effect of silver nanoparticles on soil denitrification kinetics. *Industrial Biotechnology, 8*(6), 358–364.

Vigneshwaran, N., Kathe, A. A., Varadarajan, P. V., Nachane, R. P., & Balasubramanya, R. H. (2006). Biomimetics of silver nanoparticles by white rot fungus, *Phaenerochaete chrysosporium. Colloids and Surfaces B: Biointerfaces, 53*(1), 55–59.

Vitali, F., Raio, A., Sebastiani, F., Cherubini, P., Cavalieri, D., & Cocozza, C. (2019). Environmental pollution effects on plant microbiota: The case study of poplar bacterial-fungal response to silver nanoparticles. *Applied Microbiology and Biotechnology, 103*(19), 8215–8227.

Watteau, F., & Villemin, G. (2018). Soil microstructures examined through transmission electron microscopy reveal soil-microorganisms interactions. *Frontiers in Environmental Science, 6*, 33.

Wilpiszeski, R. L., Aufrecht, J. A., Retterer, S. T., Sullivan, M. B., Graham, D. E., & Elias, D. A. (2019). Soil aggregate microbial communities: Towards understanding microbiome interactions at biologically relevant scales. *Applied and Environmental Microbiology, 85*(4).

Wolińska, A., Stępniewska, Z., & Pytlak, A. (2015). The effect of environmental factors on total soil DNA content and dehydrogenase activity. *Archives of Biological Sciences, 67*(2), 493–501.

Yadav, R. K., Singh, N. B., Singh, A., Yadav, V., Bano, C., Khare, S., & Niharika. (2020). Expanding the horizons of nanotechnology in agriculture: Recent advances, challenges and future perspectives. *Vegetos, 33*(1), 203–221.

Young, I. M., Crawford, J. W., Nunan, N., Otten, W., & Spiers, A. (2008). Microbial distribution in soils: Physics and scaling. In *Advances in Agronomy* (pp. 81–121). Academic Press.

Zhai, Y., Hunting, E. R., Wouters, M., Peijnenburg, W. J. G. M., & Vijver, M. G. (2016). Silver nanoparticles, ions, and shape governing soil microbial functional diversity: Nano shapes micro. *Frontiers in Microbiology, 7*, 1663.

Zhang, X., Xu, Z., Wu, M., Qian, X., Lin, D., Zhang, H., & Li, L. (2019). Potential environmental risks of nanopesticides: Application of $Cu(OH)_2$ nanopesticides to soil mitigates the degradation of neonicotinoid thiacloprid. *Environment International, 129*, 42–50.

Zhu, Y. G., Johnson, T. A., Su, J. Q., Qiao, M., Guo, G. X., Stedtfeld, R. D., & Tiedje, J. M. (2013). Diverse and abundant antibiotic resistance genes in Chinese swine farms. *Proceedings of the National Academy of Sciences, 110*(12), 3435–3440.

Zumft, W. G. (1997). Cell biology and molecular basis of denitrification. *Microbiology and Molecular Biology Reviews, 61*(4), 533–616.

CHAPTER 4

Configuration and Preparation Methods of Nanopesticides

VARAD NAGAR,[1] ANUJ SHARMA,[1] MAHIPAL SINGH SANKHLA,[2] KAMAKSHI PANDEY,[3,4] and GARIMA AWASTHI[2]

[1]*Department of Forensic Science, Vivekananda Global University, Jaipur, Rajasthan, India*

[2]*K. R. Mangalam University, Gurugram, Haryana, India*

[3]*Materials Research Center, Malaviya National Institute of Technology, Jaipur, Rajasthan, India*

[4]*Department of Physics, Malaviya National Institute of Technology, Jaipur, Rajasthan, India*

ABSTRACT

Nanopesticides, a new class of pesticides with nanoscale dimensions, have emerged as a promising alternative to traditional pesticides due to their enhanced efficacy and reduced environmental impact. However, the configuration and preparation of these nanopesticides are crucial factors that determine their efficacy, stability, and safety. This abstract provides an overview of the configuration and preparation of nanopesticides, including the selection of appropriate nanomaterials, fabrication techniques, and surface modifications. Various methods for characterizing nanopesticides, such as particle size, zeta potential, and morphology, are also discussed. Furthermore, the toxicity and environmental impact of nanopesticides are evaluated, and the regulatory framework governing

Nanopesticides: A Perspective from Welfare to Risk Assessment.
Garima Awasthi, Payal Lodha, and Gajendra Pal Singh (Eds.)

DOI: 10.1201/9781003605874-4

their use is highlighted. The abstract concludes by emphasizing the importance of developing efficient and sustainable methods for the configuration and preparation of nanopesticides to ensure their safe and effective use in agriculture.

4.1 INTRODUCTION

Nanopesticides represent a form of pesticide that exploits nanotechnology to enhance their effectiveness and reduce their environmental impact (Manjunatha et al., 2016). Specifically, these pesticides are designed using nanoparticles, which are particles that measure between 1 and 100 nanometers in size (Hagendorfer et al., 2010). Such nanoparticles possess unique properties that can be leveraged to create targeted and highly efficient delivery systems for pesticides (Yadav et al., 2015). For example, nanoparticles can be engineered to carry and release pesticides in a controlled manner, enabling the direct delivery of these agents to targeted pests or plants. This reduces the amount of pesticide required and limits the likelihood of harmful off-target effects on the environment (Esvelt et al., 2014). Nanopesticides offer several advantages over traditional pesticides, such as increased efficiency, reduced toxicity, and a smaller environmental footprint (Vasseghian et al., 2022). For instance, nanopesticides can be designed to release the pesticide gradually over time, reducing the need for multiple applications and decreasing the overall amount of pesticide used. Additionally, nanopesticides can be engineered to be more selective, thereby minimizing the risk of harm to non-target organisms. The use of nanopesticides in farming has the potential to revolutionize agricultural practices and reduce environmental impacts. Key benefits of nanopesticides in farming and nature include their ability to release pesticides slowly and in a controlled manner, leading to more efficient pesticide use (Bratovcic et al., 2021). This means less pesticide is needed overall, leading to cost savings and reduced environmental impact. By targeting the delivery of pesticides directly to pests or plants, nanopesticides minimize off-target effects on the environment (Prasad et al., 2014; Kah et al., 2013), reducing potential harm to non-target organisms such as pollinators and wildlife. Additionally, targeted protection against pests and diseases can improve crop yields and contribute to food security (Raj et al., 2021). Nanotechnology-based delivery systems also help reduce the development of resistance in pests, a major issue with

TABLE 4.1 Concise Overview of Different Nanopesticide Products and Their Effects on a Range of Targets

SL. No.	Pesticide Product	Active Ingredients	Size	Target	Concentration	Description	References
1.	Kocide® 3000, nanopesticide	Copper(II) hydroxide	38.7 ± 8.2 nm	Forage crops	For a year, 6.68 mg/L (Terrestrial mesocosm), 9 months at 35 mg/week (Wetland Mesocosm)	Limited effects on the biodiversity of terrestrial soil. Protists are more sensitive in aquatic ecosystems.	Carley et al. (2020)
2.	Thiamethoxam insecticide nanoformulation	Thiamethoxam	313.5 ± 7 nm	Algae, *Raphidocelis subcapitata*	0.06, 0.6, 6, 60, 600, and 1200 mg/L	*R. subcapitata*'s EC50 value was 56.15 mg/L.	Assalin et al. (2019)
3.	Kocide® 3000, Copper(II) hydroxide nanopesticide	Copper(II) hydroxide	~50 to 1000 nm	Lettuce, *Lactuca sativa*	0 mg/L, 1050 mg/L, 1555 mg/L	Decreased antioxidant production and overall antioxidant capacity.	Zhao, et al. (2016)
4.	Kocide® 3000, Copper(II) hydroxide nanopesticide	Copper(II) hydroxide	–	Cucumber, *Cucumis sativus*	0 mg/plant, 2.5 mg/plant, and 25 mg/plant	Significantly altered the mRNA levels of genes associated with antioxidant and detoxification processes.	Zhao, et al. (2017)
5.	Kocide® 3000, Copper(II) hydroxide nanopesticide	Copper(II) hydroxide	–	Maize, *Zea mays*	100 and 1000 mg/L	As leaf chlorophyll content decreased and biomass increased, the expression of POD1 and GST1 genes increased.	Zhao, et al. (2017)
6.	Kocide® 3000, Copper(II) hydroxide nanopesticide	Copper(II) hydroxide	–	Spinach, *Spinacia oleracea*	0.18 and 18 mg/plant	Decreased the amount of low molecular mass antioxidants in the leaves.	Zhao, et al. (2017)

TABLE 4.1 *(Continued)*

SL. No.	Pesticide Product	Active Ingredients	Size	Target	Concentration	Description	References
7.	CuPRO and Kocide nanopesticides	Nano-copper pesticides	–	Amphipod, *Leptocheirus plumulosus*	0.1, 0.25, 1.0, and 2.5 mg/L	A small amount of copper was released into the environment. Body weight showed little impact on respiration and survival.	Vignardi, et al. (2020)
8.	Nanopesticide based on botanical insecticide	Pyrethrum	260 nm	Honey bee, *Apis mellifera*	1 and 10 ng/µL	Had no impact on the morphology of digestive cells.	Oliveira, et al. (2019)
9	Copper nanopesticide	–	15.67–62.56 nm	*Anopheles stephensi*, *Aedes aegypti*, *Culex quinquefasciatus*, *Tenebrio molitor*, and other non-target organisms such as *Artemia salina*, *Artemia nauplii*, *Eudrilus eugeniae*, and *Eudrilus andrei*	25 µg/mL	Insects subjected to a dose of 25 g/mL showed significant mortality for *T. molitor*, *C. quinquefasciatus*, *A. stephensi*, and *A. aegypti* larvae. Non-target species exhibited lower toxicity. The findings demonstrated that *M. robertsii*-mediated copper nanoparticles are highly toxic to target pests but less harmful to non-target species.	Vivekanandhan, et al. (2021)
10	Cyantraniliprole hollow mesoporous silica poly(diacetone acrylamide) (CNAP-HMS-PDAAM)	–	–	Insect pests of rice	34.5, 39.0, and 69.0 g a.i./ha	CNAP-HMS-PDAAM demonstrated long-term control of *C. medinalis* (Guénee) and *C. suppressalis* (Walker), primarily due to its strong adhesion and slow-release properties on rice leaves. The nanocarriers also exhibited high biocompatibility with no significant impact on rice growth.	Gao, et al. (2020)

traditional pesticides. Nanopesticides contribute to sustainable agriculture by reducing the number of pesticides needed and minimizing their environmental impact (Shelar et al., 2023).

Purpose of writing a chapter on the configuration and preparation of nanopesticides is to provide a detailed understanding of how these pesticides are designed and manufactured (Vasseghian et al., 2022). This chapter serves as a guide for researchers, scientists, and engineers interested in developing nanopesticides for use in agriculture and pest control (Bratovcic et al., 2021; Nandini and Geetha, 2021). It covers various aspects of nanopesticide configuration and preparation, including the selection of nanoparticles, the design of the delivery system, and the formulation of the pesticide (Grillo et al., 2021). Information is provided on various techniques used for the synthesis and characterization of nanoparticles, such as sol-gel synthesis, sonochemical synthesis, and electrochemical synthesis. The chapter offers a comprehensive overview of nanopesticide configuration and preparation, aiming to promote the development of more effective and environmentally sustainable pest control solutions (Vasseghian et al., 2022; Wang et al., 2022).

4.2 CONFIGURATION OF NANOPESTICIDES

The configuration of nanopesticides is a complex and multifaceted process that involves the careful selection, engineering, and formulation of nanoscale materials to create effective pest control agents (Saleh et al., 2022). These nanopesticides are designed to target specific pests with precision and accuracy, while minimizing the impact on non-target organisms and the environment (Scott-Fordsmand et al., 2022). To produce nanopesticides, scientists use various complex techniques, such as nanoparticle synthesis, surface functionalization, and encapsulation, as shown in Table 4.1. These techniques allow the creation of stable and potent formulations, which can increase the bioavailability and potency of the active components (Tiwari et al., 2016). The selection of active ingredients is also a critical component of the configuration process and must be carefully considered to ensure optimal performance and safety (Desai, 2012). Designing and optimizing nanopesticides requires a thorough understanding of the physicochemical properties of the materials used, as well as the mechanisms of pest control and the potential risks associated with nanotechnology (Figure 4.1) (Adisa et al., 2019).

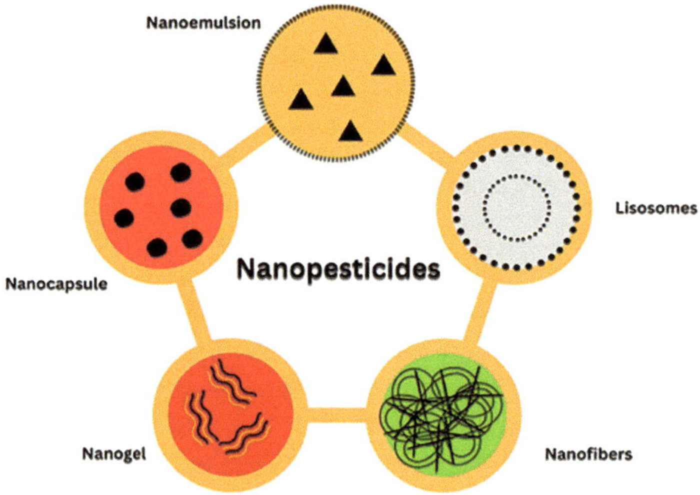

FIGURE 4.1 Different nanomaterials in the form of pesticides for crop management.

Experts in this field must possess deep knowledge of chemistry, materials science, biology, and toxicology to effectively navigate the complexities of nanopesticide development (Vasseghian et al., 2022; Abdollahdokht et al., 2022). In addition to technical expertise, innovation and refinement are crucial in the configuration of nanopesticides to ensure the safe and sustainable use of these agents in agricultural and industrial applications (Zhao et al., 2022). Ongoing research and development in this field will continue to drive advancements in nanopesticide technology, leading to more effective and sustainable pest control methods (Table 4.2) (Mittal et al., 2020).

TABLE 4.2 Configuration and Method of Utilization of Nanopesticides

Type of Nanopesticides	Size of the Particle	Mode of Action	Delivery Method	References
Neem oil nanoemulsions	Typically ranges from 20 to 200 nm	Primarily through the presence of azadirachtin, a compound in neem oil that disrupts the growth and development of insect pests.	Delivered using various methods, such as foliar spray, seed treatment, and soil drenching.	Choudhary, et al. (2022)
Essential oil nanoemulsions	Found in the range of 10 to 100 nm	Essential oils in nanoemulsions affect the nervous system, respiratory system, and other physiological processes of pests, leading to their death.	Delivered to the target pests through various methods, including foliar sprays, seed treatments, and soil drenching.	Kah, (2015)

TABLE 4.2 *(Continued)*

Type of Nanopesticides	Size of the Particle	Mode of Action	Delivery Method	References
Pyrethrin nanoemulsions	Particle size <100 nm	Pyrethrins act as neurotoxins by targeting the nervous system of insects.	Delivered through foliar spray, seed treatment, or soil drenching.	Cantó-Tejero, et al., (2022)
Spinosad nanoemulsions	Particle size 24–200 nm	Acts as a nicotinic acetylcholine receptor agonist, leading to overstimulation and disruption of the nervous system.	Spinosad nanoemulsions can be delivered via foliar spray, seed treatment, and soil drenching.	Shelar, et al. (2023)
Polymer-based nanocapsules	Particle size ranges from 10–500 nm	Nanocapsules act on target pests through modes such as contact, ingestion, and systemic uptake.	Delivered using methods like spraying, seed treatment, or soil application.	Asif, et al. (2021)
Lipid-based nanocapsules	Particle size 10–200 nm	Nanocapsules act through contact, ingestion, and systemic uptake.	Delivered via spray, seed treatment, and soil application.	Hayles, et al. (2017)
Silica-based nanocapsules	Particle size 100–200 nm	Contain insecticide imidacloprid, acting through ingestion and disrupting the nervous system.	Applied as a spray on plant leaves for aphid control.	Kashyap, et al., (2015)
Iron oxide-based nanocapsules	20–100 nm	Contain fungicide tebuconazole, acting through contact and systemic uptake by fungi.	Applied as a seed treatment for control of seed-borne fungal pathogens.	Kutawa, et al. (2021)
Hybrid nanocapsules	10–100 nm	Mode of action depends on the payload (e.g., insecticide-loaded capsules disrupt the nervous system).	Delivered via foliar spray, soil application, and seed treatment.	Chaud, et al. (2021)
Silver nanoparticles	1–100 nm	Exert pesticidal activity by disrupting cellular membranes, interfering with DNA replication and protein synthesis, and inducing oxidative stress.	Delivered via foliar sprays, seed treatments, and soil applications.	Singh, et al. (2016)
Copper nanoparticles	5–100 nm	Exert effects via membrane disruption, oxidative stress, and DNA damage.	Delivered via foliar spray, seed treatment, and soil application.	Liu, et al. (2022)

4.3 PREPARATION METHODS OF NANOPESTICIDES

Nanopesticides are a relatively new class of pesticides designed to be more efficient and targeted than traditional pesticides. They are typically synthesized using nanoparticles with pesticidal properties (Kah et al., 2018). Various methods exist for the synthesis of nanopesticides, including supramolecular chemistry, direct methods, indirect methods, and memory effect techniques. The direct method involves the precipitation of anions and cations in a pH-controlled environment (Lalia and Hashaikeh, 2021). The indirect method involves exchanging the guest solution for an anion already present in the layers. Additionally, the host-guest supramolecular complex can be created through the memory effect, which renews the molecular structure (Alarcos et al., 2017).

4.3.1 LAYERED HYDROXIDES (LH)

Hydroxide layered compounds consist of two-dimensional building blocks that are connected by weak forces. The typical formula for layered hydroxide salts (LHS) is $M^{2+}(OH)_{2}x(Am)x/m \cdot nH_2O$, and for layered double hydroxides (LDH), the formula is often $M^{2+}{}_{1}\text{-}xM^{3+}{}_{x}(OH)_2(Am)$ $x/m \cdot nH_2O$ (Zobir et al., 2021). Various counter anions can replace the positively charged inorganic layers that form the structure of layered hydroxides.

4.3.2 ZINC LAYERED HYDROXIDE

It consists of a single metal hydroxide and has a two-dimensional layered structure. It is frequently used to house different active ingredients (AI) for the manufacture of nanoinsecticides by intercalating into zinc layered hydroxide (Zn-LH) along with isoprocarb, resulting in improved thermal stability and environmental friendliness of the AI. Pesticides such as imidacloprid, 4-(2,4-dichlorophenoxy) butyrate, 2-(3-chlorophenoxy) propionate, and 2-methyl-4-chlorophenoxyacetic acid have been intercalated into Zn-LH (Zobir et al., 2021; Hashim et al., 2018; Sharif et al., 2020; Hussein et al., 2012). For the creation of nanopesticides, a pesticide called 3-glycidyloxypropyl trimethoxysilane was also intercalated into the Zn-LH host. It was found that pseudo-second-order and parabolic-diffusion models governed the behavior of the AI (Liu et al., 2018). Additionally, chlorpyrifos and sodium dodecyl sulfate were intercalated independently into the LH host, either directly (for chlorpyrifos) or indirectly (for sodium dodecyl sulfate) using a surfactant like dodecyl benzenesulfonate (Muda et al., 2020; Liu et al., 2015). Both formulations were reported to be more effective than their competitors.

4.3.3 ZINC/ALUMINUM-LAYERED DOUBLE HYDROXIDE

The host-guest supramolecular chemistry based on zinc/aluminum-layered double hydroxide (Zn/Al-LDH) has been utilized to create nanopesticides. Zn/Al-LDH was intercalated with active ingredients (AIs), such as mandelic acid, to be used as slow-release preservatives and for crop protection. Herbicides or defoliants such as 2,4-dichlorophenoxy and 4-chlorophenoxy acetate were simultaneously intercalated into Zn/Al-LDH and were found to be promising candidates for regulating the release of both AIs at controlled rates. Similarly, 2,4-dichlorophenoxy-acetic acid demonstrated improved herbicidal capabilities when glyphosate was present. Quinclorac-intercalated LDH can be used as a nanopesticide for protecting rice crops, as it is more stable than its free AI form (Sharif et al., 2020; Sharif et al., 2019). When carbonate was used as the guest, the resulting nanopesticide was shown to be superior to its counterpart as an antibacterial agent. Zn/Al-LDH-nalidixic acid nanopesticides have been employed to reduce soil and water pollution, control insect infestations, and manage plant diseases (Morales-Irigoyen et al., 2018; El-Shahawy et al., 2018). It was found that this organic biocide was highly efficient and could nearly eradicate bacteria in a short amount of time. Similar characteristics were observed for Zn/Al-LDH intercalated with pipemidic acid (Cruz et al., 2018). Additionally, 4-(2,4-Dichlorophenoxy) butyrate nanohybrids and 4-chlorophenoxyacetic acid have also been used as guests for the synthesis of nanopesticides (Bin Hussein et al., 2009; Bohari et al., 2021). In agriculture and horticulture, avermectin is used to manage insect pests on a variety of crops and ornamentals. Its intercalated product with Zn/Al-LDH exhibits a superior water-dispersible, controlled-release formulation (Zhao et al., 2011). Bacteriocinavicin provided rapid antibacterial efficacy against *Lactobacillus*, improved production, and improved stability (Fahim et al., 2017). When benzoate and its derivatives were used in the intercalation method, the release of the AI occurred slowly, with the lethality rate depending on the specific derivatives used (Mishra et al., 2019). Encouraging work has also been done on the intercalation of natural products, such as eucalyptus oil, where the AI was reduced to very low concentrations, killing *Staphylococcus* sp. and inhibiting the growth of *Pseudomonas aeruginosa* (Lobo-Sánchez et al., 2018). Lastly, pectin/cinnamate acid-intercalated Zn/Al-LDH has applications in both crop protection and cosmetics (Bugatti et al., 2019).

4.3.4 MAGNESIUM/ALUMINUM-LAYERED DOUBLE HYDROXIDE

Magnesium/aluminum-layered double hydroxide (Mg/Al-LDH) belongs to the class of LDHs where the inorganic, positively charged, two-dimensional layered structure is composed of Mg^{2+} and Al^{3+} cations. Counter anions are introduced into the interlayers to balance the charge and produce neutral materials. The decision to use Mg/Al-LDH is heavily influenced by the intended application and the advantages offered by the nanomaterial's properties, such as controlled release, enhanced thermal stability, and so forth. Nanoherbicides with slow-release properties were developed using herbicides like (4-chloro-2-methylphenoxy) acetic acid, 2,4-dichlorophenoxyacetic acid, 2-methyl-4-chlorophenoxyacetic acid, and atrazine, which help reduce environmental risks (Bruna et al., 2009; Nadiminti et al., 2019; Rebitski et al., 2019; Touloupakis et al., 2011). Imazamox, another pesticide used to stop the emergence of vegetation and plant regeneration, may lead to plant resistance with prolonged use due to its consistent mechanism of action (Authority, 2015). Although it is generally harmless to non-target species, including humans, imazamox can be hazardous by interfering with other metabolic processes. The intercalation of imazamox into Mg/Al-LDH reduces the risk of water contamination while maintaining its effectiveness and soil compatibility.

4.3.5 OTHER LAYERED DOUBLE HYDROXIDES

Other layered double hydroxides (LDHs), such as Co/Cr-, Mg/Fe-, Mg/Al/Fe-, Zn/Cr-, and Zn/Cu-LDH, have also been employed as hosts for AIs, in addition to Zn/Al-LDH and Mg/Al-LDH. For example, Co/Cr-LDH was used to intercalate haloxyfop and hexaflumuron, creating nanoinsecticides (Asiabi et al., 2018). Mg/Fe- and Mg/Al/Fe-LDH were intercalated with the herbicide 4-chloro-2-methylphenoxyacetic acid to reduce the herbicide's release and minimize environmental risks. In the development of nanopesticides with lower toxicity than their free-form counterparts, zinc pyrithione and copper pyrithione were employed as hosts in Zn/Cu-LDH for antifouling applications (Gutner-Hoch et al., 2019). The ortho-isomer of hydroxycinnamate intercalated with the same host showed a longer, more sustained release than other isomers (Majoni and Hossenlopp, 2014). Using the ion-exchange method, 4-chlorophenoxyacetate, 2,4-dichlorophenoxyacetic acid, and 2,4,5-trichlorophenoxyacetate were intercalated into $LiAl_2(OH)_6Cl \cdot xH_2O$ LDH (Li-Al-Cl LDH). The type of chlorophenoxyacetic acid used was found to affect the intercalation and deintercalation behavior of the AIs. Algaecides, which actively prevent algae growth or destroy it, have also been explored. Few papers exist on nanoalgaecides, but it was demonstrated that Zn/Fe-LDH, when used as a photocatalyst under visible light, could inhibit *Microcystis aeruginosa* without requiring an active ingredient (Gu et al., 2016).

4.4 ANALYTICAL CHARACTERIZATION METHODS FOR NANOPESTICIDES

4.4.1 TRANSMISSION ELECTRON MICROSCOPY (TEM)

Transmission Electron Microscopy (TEM) is one of the analytical characterization methods used for the evaluation of nanopesticides (Raj et al., 2021). This method allows the observation of the internal and external structure of nanoparticles, providing valuable information on their size, morphology, and distribution (Gordillo-Galeano et al., 2018). TEM is a high-resolution imaging technique that involves transmitting a beam of electrons through a thin sample (Cui et al., 2007). As the electrons pass through the sample, they interact with the atoms and produce an image on a screen or photographic film (Inkson, 2016). In the case of nanopesticides, the sample is typically prepared by placing a drop of the nanoparticle solution onto a thin carbon-coated copper grid, which is then dried under vacuum (Ullah et al., 2020). TEM provides information on the size and shape of nanoparticles, as well as their distribution and aggregation behavior. The technique can also reveal defects or imperfections in the nanoparticle structure. Furthermore, TEM can be used to investigate the interaction of nanopesticides with plant cells or insect tissues, which can provide insights into their mode of action and potential toxicity (Chhipa and Joshi, 2016). However, TEM is a time-consuming and expensive technique that requires specialized equipment and expertise. Additionally, the sample preparation process can be challenging, and there is a risk of artifacts or damage to the sample during handling. Therefore, it is important to use a combination of analytical techniques to fully characterize nanopesticides and assess their potential risks and benefits (Pachapur et al., 2016).

4.4.2 SCANNING ELECTRON MICROSCOPY (SEM)

Scanning electron microscopy (SEM) is a powerful analytical tool used for the characterization of nanopesticides (Venugopal and Sainadh, 2018). SEM is a type of electron microscope that uses a focused beam of electrons to create high-resolution images of the surface of a sample. For nanopesticides, SEM can provide detailed information about the size, shape, and morphology of the nanoparticles used in the formulation (Abdel-Azeem et al., 2023). This information is critical for understanding the physicochemical properties of the nanopesticide, as well as its potential interactions with the environment and biological systems (Mishra et al., 2019). One of the advantages of SEM is its ability to provide information about the surface topography of the nanoparticles. By scanning the surface with an electron beam, SEM generates high-resolution images that reveal details about the surface structure of the nanoparticles (Ul-Hamid, 2018). This information is particularly useful for understanding the interactions between the nanopesticide and other materials, such as plant surfaces or soil particles. SEM can also determine the size distribution of the nanoparticles in the formulation (Venugopal and Sainadh, 2016). By analyzing multiple images of the nanoparticles, it is possible to calculate the average particle size and the distribution of sizes within the sample. Additionally, SEM, combined with energy-dispersive X-ray spectroscopy (EDS), can analyze the elemental composition of nanoparticles, confirming their identity and identifying any impurities or contaminants (Utsunomiya et al., 2003).

4.4.3 DYNAMIC LIGHT SCATTERING (DLS)

Dynamic light scattering (DLS) is an analytical characterization method used for nanopesticides. It is a non-invasive and non-destructive technique that measures the size distribution of nanoparticles in a liquid suspension (Dasarahalli-Huligowda et al., 2021). DLS measures the fluctuation of light scattered by nanoparticles as they move in solution due to Brownian motion. The intensity of scattered light is directly proportional to the size of the particles (Ramos, 2017). DLS provides data on the size distribution of particles, allowing for the calculation of the mean particle size and polydispersity index (PDI), a measure of uniformity in particle size distribution (Lim et al., 2013). DLS is valuable for characterizing nanopesticides because it allows for real-time measurement of size, size distribution, and PDI, which are critical for determining their efficacy and safety in biological systems (Xu, 2015). However, DLS has limitations: it can only measure particles in the range of 1–1000 nm and cannot differentiate between particle types. Therefore, DLS is often used alongside other analytical methods, such as TEM and X-ray diffraction (XRD), for comprehensive nanopesticide characterization (Kamran et al., 2020).

4.4.4 X-RAY DIFFRACTION (XRD)

X-ray diffraction (XRD) is an analytical technique used to characterize the crystal structure of materials, including nanopesticides. XRD operates on the principle that when X-rays are directed at a material, they are diffracted by the atoms, and the resulting diffraction pattern reveals the material's crystal structure (Bunaciu et al., 2015). In the case of nanopesticides, XRD can be used to determine the crystal structure, size, and degree of crystallinity of nanoparticles. This information is important for understanding properties such as solubility, stability, and reactivity (Bradford et al., 2022). XRD can also identify impurities or other phases in the nanopesticide sample, which can affect its efficacy and safety. Additionally, XRD monitors changes in the crystal structure or size of nanoparticles over time, which is crucial for assessing the product's stability and shelf-life. XRD is a powerful tool for characterizing nanopesticides and provides valuable information for optimizing these products (Nagajyothi et al., 2020).

4.4.5 FOURIER TRANSFORM INFRARED SPECTROSCOPY (FTIR)

Fourier Transform Infrared Spectroscopy (FTIR) is a valuable analytical technique for the characterization of nanopesticides. It is a non-destructive technique that provides information on the chemical composition, molecular structure, and functional groups in a sample. FTIR measures the absorption or transmission of infrared radiation through a sample (Hsu, 1997). When infrared radiation passes through a sample, the chemical bonds absorb the radiation at specific frequencies, which are characteristic of the functional groups present in the sample. In nanopesticides, FTIR can identify the active ingredient, as well as components such as stabilizers, surfactants, or coatings (Chawla et al., 2018). FTIR can also be used to study the interaction between nanopesticides and target organisms or the environment by analyzing changes in the spectra before and after exposure (Tarhan et al., 2021). One advantage of FTIR is its ability to provide quantitative information. By comparing the absorbance values to known standards, the concentration of the active ingredient can be determined, which is useful for assessing the efficacy of the nanopesticide and ensuring production consistency (Xu et al., 2021). Overall, FTIR is a valuable analytical technique for nanopesticide characterization, providing insights into their composition, structure, and interactions, which support their development and optimization for pest management.

4.5 PHYSIO-CHEMICAL CHARACTERIZATION METHODS

4.5.1 PARTICLE SIZE AND SIZE DISTRIBUTION

Particle size and size distribution are important physicochemical properties of nanopesticides that greatly influence their behavior and effectiveness. Several techniques are available for characterizing particle size and distribution, including DLS, laser diffraction, and electron microscopy (Boyd et al., 2011). DLS is widely used for measuring particle size and distribution in nanopesticides. DLS measures fluctuations in the intensity of scattered light from particles in solution. The intensity is related to the size of the particles (Krause et al., 2010). By analyzing the correlation function of the intensity fluctuations, the particle size distribution can be determined. Laser diffraction is another popular method for determining particle size distribution, working by measuring the scattering pattern of a laser beam as it passes through a sample (Vaezifar et al., 2013). By examining the scattering pattern, particle size distribution can be estimated, as the intensity of scattered light is inversely proportional to particle size. Electron microscopy techniques like TEM and SEM are also potent methods for observing and characterizing nanopesticides (Caputo et al., 2019). These techniques allow direct visualization of particle morphology and size. In addition, electron microscopy can provide information about particle distribution within a sample and help analyze particle aggregation (Anderson et al., 2013). The characterization of particle size and distribution is crucial for the development and optimization of nanopesticides, as these properties significantly influence their behavior and effectiveness (Tipu et al., 2021).

4.5.2 ZETA POTENTIAL

Zeta potential is a physicochemical characterization method used to measure the surface charge of nanoparticles, including nanopesticides. It is a critical parameter that affects the stability, aggregation, and interaction of nanoparticles with their environment. Zeta potential is the electric potential difference between the nanoparticle surface and the surrounding liquid medium (Honary and Zahir, 2013). Several factors, including pH, ionic strength, medium composition, and surface chemistry, influence zeta potential. To measure zeta potential, a sample of nanopesticide is dispersed in a liquid medium and subjected to an electric field (Lowry et al., 2016). The movement of the nanoparticles in response to the electric field is measured to calculate the zeta potential. Nanoparticles with high absolute zeta potential values (either highly positive or highly negative) are considered stable because repulsive forces between particles prevent aggregation (Kamble et al., 2022). In contrast, nanoparticles with low zeta potential values tend to aggregate due to attractive forces between particles. Zeta potential also provides insights into the interaction of nanoparticles with target organisms (Chwalibog et al., 2010). For instance, if the zeta potential of a nanopesticide is highly positive, it may adhere more effectively to negatively charged surfaces on target organisms, improving efficacy (Vogel et al., 2017). Optimizing zeta potential can enhance the stability, efficacy, and safety of nanopesticides.

4.5.3 SURFACE AREA AND POROSITY

Physicochemical characterization methods such as surface area and porosity measurements are important tools for analyzing nanopesticides.

These techniques provide information about the physical characteristics of nanoparticles, which influence their behavior and efficiency in pest control (De Oliveira et al., 2018). Surface area measurements assess the total area available for interaction with target organisms or the environment. The Brunauer–Emmett–Teller (BET) method is commonly used for this purpose (Naderi, 2015). BET measures the quantity of gas adsorbed onto the surface of nanoparticles at different pressures, calculating the total surface area (Mohan et al., 2020). Porosity, the void space within nanoparticles, can influence their ability to absorb or release moisture, nutrients, or chemicals. Techniques like mercury intrusion porosimetry or gas adsorption are used to measure porosity and pore size distribution (Cetinbas et al., 2017). Surface area and porosity studies are valuable for optimizing nanopesticide efficacy and efficiency in pest management (Zhao et al., 2019).

4.5.4 CHEMICAL COMPOSITION AND STABILITY

Physicochemical characterization methods are critical for understanding the chemical composition and stability of nanopesticides. Techniques like differential scanning calorimetry (DSC) and X-ray diffraction (XRD) are frequently used for this purpose (Bunjes and Unruh, 2007). XRD can determine the crystallinity and crystal structure of nanopesticides, offering insights into their stability (Leyva-Porras et al., 2019). DSC assesses thermal properties such as melting point, glass transition temperature, and thermal stability. By measuring the heat flow required to maintain a temperature differential between a sample and a reference material, DSC provides information about nanopesticide stability under varying temperature and humidity conditions. Zeta potential, UV-Visible spectroscopy, and DLS are additional techniques used for physicochemical characterization (Rasheed et al., 2018). These techniques can analyze particle size distribution, electrostatic charge, and light absorption/transmission, providing information about chemical composition and stability. Physicochemical characterization methods are essential for developing and optimizing nanopesticides, ensuring their effectiveness and safety in pest management (Barazandeh et al., 2021).

4.6 CONCLUSION

In conclusion, the development of nanopesticides shows great promise as a solution to the challenges posed by conventional pesticides. Various preparation and configuration methods have been proposed, each with its advantages and limitations. The use of nanopesticides offers numerous benefits, including increased efficacy, reduced environmental impact, and improved target specificity. However, nanopesticides may also pose risks, such as toxicity and unintended effects on non-target organisms, which must be thoroughly evaluated before their widespread adoption. Further research is needed to optimize nanopesticide preparation and configuration, and to assess their long-term environmental and health impacts. With careful consideration of these factors, nanopesticides could become a valuable tool for sustainable pest control, promoting agriculture and food security.

KEYWORDS

- **agriculture**
- **nanopesticides**
- **nanoscale dimension**
- **physicochemical**
- **sustainable agriculture**
- **zeta potential**

REFERENCES

Abdel-Azeem, A. M., et al. (2023). Comparative toxicological evaluations of novel forms nanopesticides in liver and lung of albino rats. *Toxicology Reports*, 1–16.

Abdollahdokht, D., et al. (2022). Conventional agrochemicals towards nano-biopesticides: An overview on recent advances. *Frontiers in Sustainable Food Systems, 9*(1), 1–19.

Adisa, I. O., et al. (2019). Recent advances in nano-enabled fertilizers and pesticides: A critical review of mechanisms of action. *Environmental Science and Pollution Research, 6*(7), 2002–2030.

Alarcos, N., et al. (2017). Photochemistry and photophysics in silica-based materials: Ultrafast and single molecule spectroscopy observation. *The Journal of Physical Chemistry C, 117*(22), 13639–13720.

Anderson, W., et al. (2013). A comparative study of submicron particle sizing platforms: Accuracy, precision, and resolution analysis of polydisperse particle size distributions. *Journal of Nanoparticle Research, 405*, 322–330.

Asiabi, H., Yamini, Y., & Shamsayei, M. J. N. J. O. C. (2018). Using cobalt/chromium layered double hydroxide nano-sheets as a novel packed in-tube solid phase microextraction sorbent for facile extraction of acidic pesticides from water samples. *New Journal of Chemistry, 42*(12), 9935–9944.

Asif, M., et al. (2021). Nanopesticides application in agriculture and their impact on environment. In *Agricultural Development in Asia: Potential Use of Nano-Materials and Nano-Technology*. IntechOpen.

Assalin, M. R., et al. (2019). Nanoformulation as a tool for improvement of thiamethoxam encapsulation and evaluation of ecotoxicological impacts. *Energy, Ecology and Environment, 4*, 310–317.

Authority, E. J. E. J. (2015). Conclusion on the peer review of the pesticide risk assessment of the active substance ferric phosphate. *EFSA Journal, 13*(1).

Barazandeh, A., et al. (2021). Spectrophotometric determination of naproxen using chitosan capped silver nanoparticles in pharmaceutical formulation. *Journal of Pharmaceutical Analysis, 34*(7), 1576–1585.

Bin Hussein, M. Z., et al. (2009). Controlled release formulation of agrochemical pesticide based on 4-(2, 4-dichlorophenoxy) butyrate nanohybrid. *Journal of Nanomaterials, 9*(3), 2140–2147.

Bohari, F. L., et al. (2021). The synthesis and characterization of layered hydroxide salt interleaved with 4-chlorophenoxyacetic acid and its control release study as herbicide. In *AIP Conference Proceedings* (AIP Publishing LLC).

Boyd, R. D., et al. (2011). New approach to inter-technique comparisons for nanoparticle size measurements; using atomic force microscopy, nanoparticle tracking analysis and dynamic light scattering. *Journal of Nanoparticle Research, 387*(1–3), 35–42.

Bradford, S. A., et al. (2022). Environmental applications and risks of nanomaterials: An introduction to CREST publications during 2018–2021. *Environmental Engineering Journal, 52*(21), 3753–3762.

Bratovcic, A., et al. (2021). Nanopesticides and nanofertilizers and agricultural development: Scopes, advances and applications. *Agronomy, 11*(4), 301–316.

Bruna, F., et al. (2009). Layered double hydroxides as adsorbents and carriers of the herbicide (4-chloro-2-methylphenoxy) acetic acid (MCPA): Systems Mg–Al, Mg–Fe, and Mg–Al–Fe. *Applied Clay Science, 168*(2, 3), 1476–1481.

Bugatti, V., et al. (2019). Green pesticides based on cinnamate anion incorporated in layered double hydroxides and dispersed in pectin matrix. *Journal of Environmental Chemical Engineering, 209*, 356–362.

Bunaciu, A. A., Udriştioiu, E. G., & Aboul-Enein, H. Y. J. C. R. I. A. C. (2015). X-ray diffraction: Instrumentation and applications. *Critical Reviews in Analytical Chemistry, 45*(4), 289–299.

Bunjes, H., & Unruh, T. J. A. D. D. R. (2007). Characterization of lipid nanoparticles by differential scanning calorimetry, X-ray and neutron scattering. *Advanced Drug Delivery Reviews, 59*(6), 379–402.

Cantó-Tejero, M., Guirao, P., & Pascual-Villalobos, M. J. J. P. M. S. (2022). Aphicidal activity of farnesol against the green peach aphid—Myzus persicae. *Pest Management Science, 78*(6), 2714–2721.

Caputo, F., et al. (2019). Measuring particle size distribution of nanoparticle-enabled medicinal products, the joint view of EUNCL and NCI-NCL. A step-by-step approach combining orthogonal measurements with increasing complexity. *European Journal of Pharmaceutical Sciences, 299*, 31–43.

Carley, L. N., et al. (2020). Long-term effects of copper nanopesticides on soil and sediment community diversity in two outdoor mesocosm experiments. *Environmental Science & Technology, 54*(14), 8878–8889.

Cetinbas, F. C., et al. (2017). Hybrid approach combining multiple characterization techniques and simulations for microstructural analysis of proton exchange membrane fuel cell electrodes. *International Journal of Hydrogen Energy, 344*, 62–73.

Chaud, M., et al. (2021). Nanopesticides in agriculture: Benefits and challenges in agricultural productivity, toxicological risks to human health and environment. *Nanomaterials, 9*(6), 131.

Chawla, P., et al. (2018). Organophosphorus pesticides residues in food and their colorimetric detection. *Journal of Environmental Sciences, 10*, 292–307.

Chhipa, H., Joshi, P. J. N. I. F. (2016). Nanofertilizers, nanopesticides and nanosensors in agriculture. *Nanotechnology and Food Safety*, 247–282.

Choudhary, P., et al. (2022). Synthesis and evaluation of naringin-loaded neem oil nanopesticidal emulsion for sustainable crop management system. *Journal of Agricultural and Food Chemistry*.

Chwalibog, A., et al. (2010). Visualization of interaction between inorganic nanoparticles and bacteria or fungi. *Journal of Nanoparticle Research, 12*, 1085–1094.

Cruz, A. S., et al. (2018). Organic biocides hosted in layered double hydroxides: Enhancing antimicrobial activity. *Green Chemistry, 16*(1), 163–169.

Cui, H., et al. (2007). Elucidating the assembled structure of amphiphiles in solution via cryogenic transmission electron microscopy. *Langmuir, 3*(8), 945–955.

Dasarahalli-Huligowda, L. K., et al. (2021). Role of nanotechnology: Emerging path from soil to fork. *In Advances in Food and Nutrition Research* (pp. 3–30).

De Oliveira, J. L., et al. (2018). Geraniol encapsulated in chitosan/gum arabic nanoparticles: A promising system for pest management in sustainable agriculture. *Industrial Crops and Products, 66*, 5325–5334.

Desai, N. J. T. A. J. (2012). Challenges in development of nanoparticle-based therapeutics. *Therapeutic Advances in Medical Oncology, 14*(2), 282–295.

El-Shahawy, A. A., et al. (2018). Synthesis and evaluation of layered double hydroxide/doxycycline and cobalt ferrite/chitosan nanohybrid efficacy on gram positive and gram negative bacteria. *Journal of the Science of Food and Agriculture, 91*, 361–371.

Esvelt, K. M., et al. (2014). Concerning RNA-guided gene drives for the alteration of wild populations. *eLife, 3*, e03401.

Fahim, H. A., et al. (2017). Enhancement of the productivity of the potent bacteriocin avicin A and improvement of its stability using nanotechnology approaches. *Scientific Reports, 7*(1), 10604.

Gao, Y., et al. (2020). Efficacy of an adhesive nanopesticide on insect pests of rice in field trials. *Journal of Asia-Pacific Entomology, 23*(4), 1222–1227.

Gordillo-Galeano, A., Mora-Huertas, C. E. J. E. J. O. P., & Biopharmaceutics. (2018). Solid lipid nanoparticles and nanostructured lipid carriers: A review emphasizing on particle structure and drug release. *European Journal of Pharmaceutical Sciences, 133*, 285–308.

Grillo, R., et al. (2021). Ecotoxicological and regulatory aspects of environmental sustainability of nanopesticides. *Environmental Pollution, 404*, 124148.

Gu, N., et al. (2016). Microcystis aeruginosa inhibition by Zn–Fe–LDHs as photocatalyst under visible light. *Journal of Photochemistry and Photobiology A: Chemistry, 64*, 189–195.

Gutner-Hoch, E., et al. (2019). Toxicity of engineered micro-and nanomaterials with antifouling properties to the brine shrimp Artemia salina and embryonic stages of the sea urchin Paracentrotus lividus. *Journal of Hazardous Materials, 251*, 530–537.

Hagendorfer, H., et al. (2010). Size-fractionated characterization and quantification of nanoparticle release rates from a consumer spray product containing engineered nanoparticles. *Environmental Science: Nano, 12*, 2481–2494.

Hashim, N., et al. (2018). The effect of ion exchange and co-precipitation methods on the intercalation of 3-(4-methoxyphenyl) propionic acid into layered zinc hydroxide nitrate. *Chemical Engineering Journal, 25*, 249–258.

Hayles, J., et al. (2017). Nanopesticides: A review of current research and perspectives. *Journal of Nanopesticides, 193–225*.

Honary, S., & Zahir, F. J. T. J. O. P. R. (2013). Effect of zeta potential on the properties of nano-drug delivery systems: A review (Part 1). *Journal of Pharmaceutical Research, 12*(2), 255–264.

Hsu, C. P. S. J. H. O. I. T. F. A. C. (1997). Infrared spectroscopy. *Handbook of Infrared Spectroscopy of Ultrathin Films*, 249.

Hussein, M. Z., et al. (2012). Synthesis of a monophasic nanohybrid for a controlled release formulation of two active agents simultaneously. *Materials Chemistry and Physics, 58*, 60–66.

Inkson, B. J. (2016). Scanning electron microscopy (SEM) and transmission electron microscopy (TEM) for materials characterization. In *Materials Characterization Using Nondestructive Evaluation (NDE) Methods* (pp. 17–43). Elsevier.

Kah, M. J. (2015). Nanopesticides and nanofertilizers: Emerging contaminants or opportunities for risk mitigation? *Frontiers in Chemistry, 3*, 64.

Kah, M., et al. (2013). Nanopesticides: State of knowledge, environmental fate, and exposure modeling. *Journal of Environmental Quality*.

Kah, M., et al. (2018). A critical evaluation of nanopesticides and nanofertilizers against their conventional analogues. *Journal of Environmental Science and Health, Part B, 13*(8), 677–684.

Kamble, S., et al. (2022). Revisiting zeta potential, the key feature of interfacial phenomena, with applications and recent advancements. *Journal of Nanobiotechnology, 7*(1), e202103084.

Kamran, K., et al. (2020). Nano zinc elicited biochemical characterization, nutritional assessment, antioxidant enzymes, and fatty acid profiling of rapeseed. *Journal of Nanobiotechnology, 15*(11), e0241568.

Kashyap, P. L., Xiang, X., & Heiden, P. J. I. J. O. B. M. (2015). Chitosan nanoparticle based delivery systems for sustainable agriculture. *Journal of Biochemical and Molecular Toxicology, 77*, 36–51.

Krause, B., et al. (2010). Dispersability and particle size distribution of CNTs in an aqueous surfactant dispersion as a function of ultrasonic treatment time. *Nanotechnology, 48*(10), 2746–2754.

Kumar, P., et al. (2021). Bacillus-based nano-bioformulations for phytopathogens and insect–pest management. *Nanobiotechnology, 31*, 1–11.

Kutawa, A. B., et al. (2021). Trends in nanotechnology and its potentialities to control plant pathogenic fungi: A review. *Nanotechnology, 10*(9), 881.

Lalia, B. S., & Hashaikeh, R. J. D. (2021). Electrochemical precipitation to reduce waste brine salinity. *Separation and Purification Technology, 498*, 114796.

Leyva-Porras, C., et al. (2019). Application of differential scanning calorimetry (DSC) and modulated differential scanning calorimetry (MDSC) in food and drug industries. *Molecules, 12*(1), 5.

Lim, J., et al. (2013). Characterization of magnetic nanoparticle by dynamic light scattering. *Journal of Nanotechnology, 8*, 1–14.

Liu, J., et al. (2015). Preparation and release behavior of chlorpyrifos adsolubilized into layered zinc hydroxide nitrate intercalated with dodecylbenzenesulfonate. *Journal of Agricultural and Food Chemistry, 7*(21), 11180–11188.

Liu, J., et al. (2018). Intercalation, characterization and release behavior of imidacloprid into layered hydroxide salts by coupling of (3-glycidyloxypropyl) trimethoxysilane. *Journal of Hazardous Materials, 553*, 42–49.

Liu, P., et al. (2022). Efficacy-associated cost analysis of copper-based nanopesticides for tomato disease control. *Nanomaterials, 2*(4), 796–804.

Lobo-Sánchez, M., et al. (2018). ZnAl layered double hydroxides impregnated with eucalyptus oil as efficient hybrid materials against multi-resistant bacteria. *Applied Clay Science, 153*, 61–69.

Lowry, G. V., et al. (2016). Guidance to improve the scientific value of zeta-potential measurements in nanoEHS. *Nature Nanotechnology, 3*(5), 953–965.

Majoni, S., & Hossenlopp, J. M. J. A. I. P. C. (2014). Controlled release kinetics in hydroxy double salts: Effect of host anion structure. *AIP Conference Proceedings*.

Manjunatha, S., Biradar, D., & Aladakatti, Y. R. J. J. F. S. (2016). Nanotechnology and its applications in agriculture: A review. *Journal of Agricultural Science and Food Science, 29*(1), 1–13.

Mishra, G., Dash, B., & Pandey, S. J. A. C. S. (2019). Effect of molecular dimension on gallery height, release kinetics and antibacterial activity of ZnAl layered double hydroxide (LDH) encapsulated with benzoate and its derivatives. *Applied Clay Science, 181*, 105230.

Mishra, M., et al. (2019). Prospects, challenges and need for regulation of nanotechnology with special reference to India. *Environmental Toxicology and Pharmacology, 171*, 677–682.

Mittal, D., et al. (2020). Nanoparticle-based sustainable agriculture and food science: Recent advances and future outlook. *Frontiers in Sustainable Food Systems, 2*, 579954.
Mohan, V. B., Jayaraman, K., & Bhattacharyya, D. J. S. S. C. (2020). Brunauer–Emmett–Teller (BET) specific surface area analysis of different graphene materials: A comparison to their structural regularity and electrical properties. *Carbon, 320*, 114004.
Morales-Irigoyen, E. E., et al. (2018). A bionanohybrid ZnAl-NADS ecological pesticide as a treatment for soft rot disease in potato (*Solanum tuberosum* L.). *Materials Letters, 25*(22), 21430–21439.
Muda, Z., et al. (2020). Adsolubilisation of thiacloprid pesticide into the layered zinc hydroxide salt intercalated with dodecyl sulfate, for controlled release formulation. *Journal of Environmental Management, 24*(5), 279–288.
Naderi, M. (2015). Surface area: Brunauer–Emmett–Teller (BET). In *Progress in Filtration and Separation* (pp. 585–608). Elsevier.
Nadiminti, P. P., et al. (2019). Use of Mg–Al nanoclay as an efficient vehicle for the delivery of the herbicide 2,4-dichlorophenoxyacetic acid. *Journal of Nanoscience and Nanotechnology, 7*(12), 10962–10970.
Nagajyothi, P. C., et al. (2020). Green synthesis: Photocatalytic degradation of textile dyes using metal and metal oxide nanoparticles—Latest trends and advancements. *Environmental Science and Pollution Research, 50*(24), 2617–2723.
Nandini, B., & Geetha, N. (2021). Smart delivery mechanisms of nanofertilizers and nanocides in crop biotechnology. In *Advances in Nano-Fertilizers and Nanopesticides in Agriculture* (pp. 385–414). Elsevier.
Oliveira, C. R., et al. (2019). Nanopesticide based on botanical insecticide pyrethrum and its potential effects on honeybees. *Chemosphere, 236*, 124282.
Pachapur, V. L., et al. (2016). Behavior and characterization of titanium dioxide and silver nanoparticles in soils. *Environmental Science and Pollution Research, 563*, 933–943.
Prasad, R., Kumar, V., & Prasad, K. S. J. A. J. O. B. (2014). Nanotechnology in sustainable agriculture: Present concerns and future aspects. *Journal of Biofertilizers and Biopesticides, 13*(6), 705–713.
Raj, S. N., et al. (2021). A comprehensive review on regulatory invention of nanopesticides in agricultural nano formulation and food system. *Agricultural Sciences, 1239*, 130517.
Ramos, A. P. (2017). Dynamic light scattering applied to nanoparticle characterization. In *Nanocharacterization Techniques* (pp. 99–110). Elsevier.
Rasheed, T., et al. (2018). Catalytic potential of bio-synthesized silver nanoparticles using Convolvulus arvensis extract for the degradation of environmental pollutants. *Journal of Environmental Management, 181*, 44–52.
Rebitski, E. P., Darder, M., & Aranda, P. J. B. J. (2019). Layered double hydroxide/sepiolite hybrid nanoarchitectures for the controlled release of herbicides. *Journal of Nanoparticle Research, 10*(1), 1679–1690.
Saleh, M. M., et al. (2022). Nanotechnological approaches for efficient delivery of plant ingredients. In *Sustainable Agriculture Reviews 53: Nanoparticles: A New Tool to Enhance Stress Tolerance* (pp. 247–286). Springer.
Scott-Fordsmand, J., Fraceto, L., & Amorim, M. J. J. (2022). Nanopesticides: The lunch-box principle—Deadly goodies (semio-chemical functionalized nanoparticles that deliver pesticide only to target species). *Journal of Nanopesticides, 20*(1), 1–9.

Sharif, S. N. M., et al. (2019). Preparation and characterization of novel paddy cultivation herbicide nanocomposite from zinc/aluminum layered double hydroxide and quinclorac anion. *Journal of Environmental Science and Health, Part B, 23*(5), 260–265.

Sharif, S. N. M., et al. (2020). The influence of chitosan coating on the controlled release behavior of zinc/aluminum-layered double hydroxide-quinclorac composite. *Journal of Hazardous Materials, 251*, 123076.

Shelar, A., et al. (2023). Recent advances in nano-enabled seed treatment strategies for sustainable agriculture: Challenges, risk assessment, and future perspectives. *Frontiers in Plant Science, 15*(1), 54.

Singh, P., et al. (2016). Biological synthesis of nanoparticles from plants and microorganisms. *Journal of Plant Biology, 34*(7), 588–599.

Širc, J., et al. (2012). Morphological characterization of nanofibers: Methods and application in practice. *Nanofibers, 2012*, 1–14.

Tarhan, O., Spotti, M. J. J. C., & Biointerfaces, S. B. (2021). Nutraceutical delivery through nano-emulsions: General aspects, recent applications, and patented inventions. *Colloids and Surfaces B: Biointerfaces, 200*, 111526.

Tipu, M. M. H., et al. (2021). Potential applications of nanotechnology in agriculture: A smart tool for sustainable agriculture. In *Agricultural Development in Asia-Potential Use of Nano-Materials and Nano-Technology* (pp. 1–18). IntechOpen.

Tiwari, R. V., Patil, H., & Repka, M. A. J. E. O. O. D. D. (2016). Contribution of hot-melt extrusion technology to advance drug delivery in the 21st century. *Expert Opinion on Drug Delivery, 13*(3), 451–464.

Touloupakis, E., Margelou, A., & Ghanotakis, D. F. J. P. M. S. (2011). Intercalation of the herbicide atrazine in layered double hydroxides for controlled-release applications. *Pest Management Science, 67*(7), 837–841.

Ul-Hamid, A. (2018). *A Beginners' Guide to Scanning Electron Microscopy* (Vol. 1). Springer.

Ullah, H., et al. (2020). *In vivo* phytotoxicity, uptake, and translocation of PbS nanoparticles in maize (*Zea mays* L.) plants. *Journal of Hazardous Materials, 737*, 139558.

Utsunomiya, S., Ewing, R. C. J. E. S., & Technology. (2003). Application of high-angle annular dark field scanning transmission electron microscopy, scanning transmission electron microscopy-energy dispersive X-ray spectrometry, and energy-filtered transmission electron microscopy to the characterization of nanoparticles in the environment. *Environmental Science & Technology, 37*(4), 786–791.

Vaezifar, S., et al. (2013). Effects of some parameters on particle size distribution of chitosan nanoparticles prepared by ionic gelation method. *Journal of Nanoparticle Research, 24*, 891–903.

Vasseghian, Y., et al. (2022). Metal-organic framework-enabled pesticides are an emerging tool for sustainable cleaner production and environmental hazard reduction. *Environmental Pollution, 299*, 133966.

Venugopal, N., & Sainadh, N. J. I. J. O. N. (2016). Novel polymeric nanoformulation of Mancozeb—An eco-friendly nanomaterial. *International Journal of Nanotechnology, 15*(04), 1650016.

Venugopal, N., & Sainadh, N. J. M. T. P. (2018). Synthesis and characterization of Nano-tricyclazole-A systemic fungicide. *Materials Today: Proceedings, 5*(8), 16807–16813.

Vignardi, C. P., et al. (2020). Conventional and nano-copper pesticides are equally toxic to the estuarine amphipod *Leptocheirus plumulosus*. *Aquatic Toxicology, 224*, 105481.
Vivekanandhan, P., et al. (2021). Insecticidal efficacy of microbial-mediated synthesized copper nano-pesticide against insect pests and non-target organisms. *International Journal of Environmental Research and Public Health, 18*(19), 10536.
Vogel, R., et al. (2017). High-resolution single particle zeta potential characterization of biological nanoparticles using tunable resistive pulse sensing. *Scientific Reports, 7*(1), 17479.
Wang, D., et al. (2022). Nano-enabled pesticides for sustainable agriculture and global food security. *Nature Sustainability, 17*(4), 347–360.
Xu, C., et al. (2018). Emulsion-based synchronous pesticide encapsulation and surface modification of mesoporous silica nanoparticles with carboxymethyl chitosan for controlled azoxystrobin release. *Chemical Engineering Science, 348*, 244–254.
Xu, R. J. P. (2015). Light scattering: A review of particle characterization applications. *Particles, 18*, 11–21.
Xu, Y., et al. (2021). Size effect of mesoporous silica nanoparticles on pesticide loading, release, and delivery in cucumber plants. *Journal of Nanobiotechnology, 11*(2), 575.
Yadav, A., et al. (2015). Fungi as an efficient mycosystem for the synthesis of metal nanoparticles: Progress and key aspects of research. *Biotechnology Advances, 37*, 2099–2120.
Zhao, J., et al. (2011). Water dispersible avermectin-layered double hydroxide nanocomposites modified with sodium dodecyl sulfate. *Applied Clay Science, 51*(4), 460–466.
Zhao, L., et al. (2016). Application of metabolomics to assess the impact of Cu (OH)2 nanopesticide on the nutritional value of lettuce (*Lactuca sativa*): Enhanced Cu intake and reduced antioxidants. *NanoImpact, 3*, 58–66.
Zhao, L., et al. (2017). Metabolomics reveals Cu (OH)2 nanopesticide-activated anti-oxidative pathways and decreased beneficial antioxidants in spinach leaves. *Environmental Science & Technology, 51*(17), 10184–10194.
Zhao, L., et al. (2017). Response at genetic, metabolic, and physiological levels of maize (Zea mays) exposed to a Cu (OH)2 nanopesticide. *ACS Sustainable Chemistry & Engineering, 5*(9), 8294–8301.
Zhao, W., et al. (2022). Engineered Zn-based nanopesticides as an opportunity for treatment of phytopathogens in agriculture. *Journal of Cleaner Production, 28*, 100420.
Zhao, X., et al. (2017). Development strategies and prospects of nano-based smart pesticide formulation. *Journal of Agricultural and Food Chemistry, 66*(26), 6504–6512.
Zhao, X., et al. (2019). Study on pore structures of tight sandstone reservoirs based on nitrogen adsorption, high-pressure mercury intrusion, and rate-controlled mercury intrusion. *Journal of Natural Gas Science and Engineering, 141*(11).
Zobir, S. A. M., et al. (2021). A review on nanopesticides for plant protection synthesized using the supramolecular chemistry of layered hydroxide hosts. *Nanomaterials, 10*(11), 1077.

CHAPTER 5

Categories of Nanopesticide Used for Crop Protection

SHIVANGI GIRI,[1] LATA SHAHANI,[1] DEV SAINI,[2] SUJATA SHARMA,[2] and KUMUD KANT AWASTHI[3]

[1]*Department of Zoology, IIS Deemed to be University, Jaipur, Rajasthan, India*

[2]*Department of Life Sciences, Vivekananda Global University, Jaipur, Rajasthan, India*

[3]*K. R. Mangalam University, Gurugram, Haryana, India*

ABSTRACT

Increased demand for food supplies brought on by the quick population expansion has led to greater use of agrochemicals to close the gap between food production and consumption. The existing agricultural production system has a significant challenge across the world because of crop losses brought on by weeds, insects, and plant infections. Plants can be protected from a variety of pests, such as diseases and insects, by using pesticides, which are biological or chemical substances. Chemical pesticides used carelessly can cause serious health risks, environmental damage, and higher manufacturing costs. Therefore, new strategies are required to safeguard crop plants with little to no negative environmental consequences. The creation of several Nanoscale pesticide formulations has demonstrated increased pesticidal efficacy, controlled release capability, and targeted delivery features. Significant advancements in nanotechnology and its use in several sectors have been made during the past 10 years.

Nanopesticides: A Perspective from Welfare to Risk Assessment.
Garima Awasthi, Payal Lodha, and Gajendra Pal Singh (Eds.)

DOI: 10.1201/9781003605874-5

5.1 INTRODUCTION

Pesticides put a crucial role in modern agriculture, protecting crops and controlling pests that threaten food security. However, traditional pesticides come with their own set of limitations and risks, including toxicity, non-specificity, and environmental persistence. As a result, the development of nanopesticides has gained attention as a promising alternative to traditional pesticides. Nanopesticides have unique physical and chemical properties that allow for greater precision and effectiveness, lower toxicity, and reduced environmental impact (Lade et al., 2019).

The main worldwide issues include trying to increase food production while dealing with factors including population expansion, limited land and water resources, climate change, high rates of illnesses and insect pests, and agrochemicals' bioaccumulation. Significant technical developments and breakthroughs have been developed in recent years in the sector of agriculture to solve the growing issues of sustainable production and food security. In particular, nanotechnology has the ability to offer practical answers to the numerous difficulties associated to agriculture (Pandey et al., 2018). The best choice for increasing solubility, dispersion, bioavailability, preventing abortive degradation, and allowing for the controlled release of active components is nanopesticides. Nanoinsecticides provide controlled release dependent on light, pH, humidity, and temperature. In nature, volatile essential oils with insecticidal properties are vulnerable to UV radiation and break down when exposed to light. These issues will be mitigated and these essential oils will become strong candidates for pesticides when loaded into nanoparticles (NPs) (Anandhi et al., 2020).

Nanopesticides are designed to be more effective and efficient than traditional pesticides. These pesticides are made up of NPs that are engineered to deliver the active ingredient to the target pest or plant in a more controlled and targeted manner (Kah et al., 2013).

One potential benefit of using nanopesticides in crop production is that they may reduce the amount of pesticide needed to control pests and diseases. This is because the NPs are designed to deliver the active ingredient more effectively, so less pesticide is needed to achieve the same level of control. Additionally, the targeted delivery of the active ingredient may reduce the impact of the pesticide on non-target organisms and the environment (Guan et al., 2010). However, the use of nanopesticides is still a relatively new and developing technology, and there are concerns about the potential risks and unintended consequences associated with their use. For example, the NPs in nanopesticides may have different toxicological properties than larger particles of the same substance, and there are concerns about their potential impact on human health and the environment. Overall, while nanopesticides have the potential to be a valuable tool in crop production, it is important to carefully assess their risks and benefits and to continue to study their potential impacts on the environment and human health.

The formulation of nanopesticides involves preparing nanoparticles (NPs) that contain the active ingredient(s) of the pesticide. The process includes several key steps:

- *Synthesis of NPs:* NPs are synthesized using various methods, including chemical precipitation, sol-gel, microemulsion, and sonochemical methods (Abd-Elsalam et al., 2018). The size, shape, and surface properties of NPs can be controlled during synthesis to optimize their performance.
- *Loading of Active Ingredient:* The active ingredient of the pesticide is loaded into the NPs through techniques such as adsorption, encapsulation, and covalent bonding.
- *Stabilization:* The NPs are stabilized using surfactants, polymers, or other agents to prevent aggregation and maintain stability in solution (Letchford and Burt, 2007).
- *Characterization:* The NPs are characterized to determine their size, shape, surface properties, and stability.
- *Testing:* Finally, the nanopesticide formulation is tested for efficacy against pests and assessed for environmental and human health safety.

5.2 CATEGORIES OF NANOPESTICIDES

According to their intended use, nanoformulations of pesticides can be divided into three categories. Kah et al. (2013) stated that there are formulations available that can enhance the solubility of active ingredients that are not soluble in water, formulations that can decrease the speed at which the active ingredients are released, and formulations that can ensure targeted delivery and greater chemical stability.

To assess their potential impact on the environment, nanopesticides were initially classified based on their intended purpose and then divided into subcategories based on the anticipated variations in their environmental fate and the types of active ingredients used.

Nanopesticides (NPs) are a type of pesticide that utilizes NPs to improve the effectiveness and efficiency of crop protection. There are several categories of nanopesticides used for crop protection, as discussed below.

5.2.1 ON THE BASIS OF NANOPARTICLES (NPS)

There are different categories of nanopesticides based on the type of NPs used as carriers or delivery agents. Here are some examples:

1. **Metal-Based Nanopesticides:** These use NPs made of metals such as silver, copper, or zinc to deliver the active ingredients. These NPs can penetrate the pest's exoskeleton and disrupt its physiological functions, leading to death (Guan et al., 2006);
2. **Polymeric Nanopesticides:** These use NPs made of polymers to encapsulate the active ingredients (Letchford, et al., 2007). The polymer shell protects the active ingredients from degradation and increases their stability and efficacy (Salma et al., 2010);
3. **Lipid-Based Nanopesticides:** These use NPs made of lipids to encapsulate the active ingredients. The lipid shell can enhance the penetration of the active ingredients into the target pest's cells, increasing their effectiveness (Lai et al., 2006);
4. **Carbon-Based Nanopesticides:** These use NPs made of carbon, such as fullerenes or carbon nanotubes, to deliver the active ingredients. Carbon-based NPs have unique physicochemical properties that can enhance the delivery and efficacy of the active ingredients (Zhu et al., 2022);
5. **Bio-Based Nanopesticides:** These are nanopesticides that use natural or bio-based NPs, such as chitosan, as the active ingredient (Lade et al., 2019). They are environmentally friendly and can be used for organic farming (Jampílek et al., 2019; Liang et al., 2018) (Figure 5.1; Tables 5.1 and 5.2).

TABLE 5.1 Nanopesticides, Mode of Entry, Matrix, Size, and Composition of Nanopesticides

Categories of Nanopesticides	Mode of Entry	Matrix	Size	Composition of Nanopesticides	References
Type-1	Via plant-soil system	Metal-based (Ag and Cu)	1–10 nm	Ag, Ti, Zn, Fe, Al-based	Kah et al. (2013); Letchford, et al. (2007); Wang et al. (2022)
Type-2	Above ground shoot	Polymer & clay-based	10–300 nm	Polymers like chitosan, cellulose, polyethylene, clay- silica, montmorillonite, etc.	

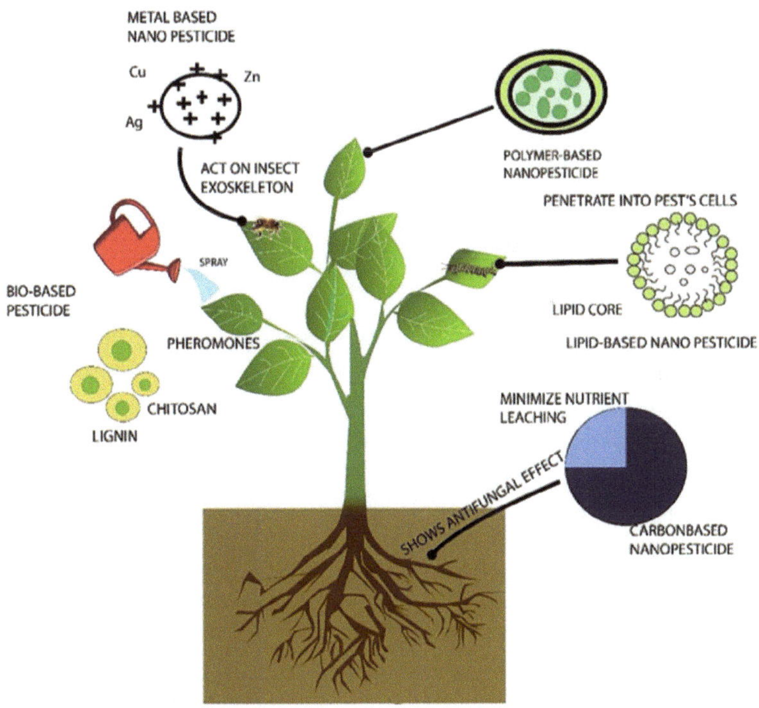

FIGURE 5.1 Category of nanopesticides based on NPs and summary of application on crop and effect on pest.

5.2.2 ON THE BASIS OF NUMBER OF FACTORS

Formulations for nanopesticides can be categorized based on a number of factors. For example, they can be categorized based on their intended use as:

TABLE 5.2 Nanopesticides and Their NPs Formulation and Its Effect on Pests

Categories of Nanopesticides	NPs Formulation	Effect on Pests	References
Metal-based nanopesticides	Copper oxide, zinc oxide, and silver NPs	Penetrate the pest's exoskeleton and disrupt its physiological functions, leading to death	Guan et al. (2006)
Lipid-based nanopesticides	Lipids to encapsulate the active ingredients	Penetration of active ingredients into the target pest's cells	Lai et al. (2006)
Polymeric nanopesticides	Polymers to encapsulate the active ingredients	Polymer shell protects the active ingredients from degradation to pest's cell and increases their stability and efficacy	Salma et al. (2010)
Carbon-based nanopesticides	Carbon or carbon nanotubes, to deliver the active ingredients	Have unique physicochemical properties it enhance the delivery and efficacy of the active ingredients	Zhu et al. (2022)
Bio-based nanopesticides	Natural or bio-based NPs, such as chitosan	Used for organic farming, are like pheromones	Liang et al. (2018); Lade et al., (2019)

1. Emulsifiable formulations that improve the solubility of pesticides that are poorly water-soluble.
2. Formulations that slow the release rate of the encapsulated pesticide.
3. Formulations capable of achieving targeted delivery and offering protection against early degradation (Kah et al., 2013).

Nanopesticide refers to a broad range of goods and cannot be regarded as standing for a single category. Several surfactants, polymers, and metal NPs in the nanometer size range are combined in several nanoformulations. The objectives of nanoformulations are often similar to those of traditional pesticide formulations; they include enhancing the apparent solubility of insufficiently soluble active chemicals, releasing the active ingredient slowly and precisely, and/or guarding against premature degradation.

5.2.3 ON THE BASIS OF FORMULATIONS

1. **Nanoemulsions:** They are one of the most widely used categories of nanopesticides due to their excellent physical and chemical

stability. They consist of an oil phase, an aqueous phase, and a surfactant that stabilizes the mixture. Nanoemulsions have a small droplet size, typically less than 100 nm, which allows for better penetration of active ingredients into pests and plant tissues (Sabry et al., 2018). Nanoemulsions can be applied through various methods, such as spraying, dipping, and seed treatment. They have shown promising results in controlling a variety of pests, including fungi, bacteria, and insects (Song et al., 2009). A pesticide nanoemulsion is a type of pesticide formulation where the pesticide is dispersed in water as nanosized droplets using surfactant molecules. The surfactant molecules help to stabilize the droplets and prevent them from coalescing, which allows the pesticide to remain in suspension and improve its effectiveness (Hossain et al., 2019). Nanoemulsions have several advantages over conventional pesticide formulations. Firstly, they can improve the solubility and bioavailability of the pesticide, which means that less pesticide is required to achieve the same level of control. Secondly, nanoemulsions can provide more uniform coverage of the target surface, which can improve the effectiveness of the pesticide. Finally, nanoemulsions can reduce the environmental impact of the pesticide by reducing the amount of active ingredient required and minimizing off-target effects (Abd-Elsalam et al., 2018).

2. **Nanocapsules:** They are another category of nanopesticides that consist of a core containing the active ingredient, surrounded by a polymeric shell (Sabry et al., 2018). The size of nanocapsules typically ranges from 10 to 1000 nm, depending on the type of polymer used. The polymeric shell provides a protective barrier that prevents the active ingredient from degrading and enhances its stability (Letchford and Burt, 2007). Nanocapsules have been used to control a wide range of pests, including aphids, mites, and whiteflies (Taylor et al., 2005).

3. **Nanoparticles (NPs):** NPs are a broad category of nanopesticides that encompass a range of particle types, including metal NPs, metal oxide NPs, and carbon-based NPs. They are generally used for their physical and chemical properties, such as high surface area, high reactivity, and controlled release of active ingredients.

Metal NPs, such as silver and copper, have shown promising results in controlling bacterial and fungal infections in crops. Metal oxide NPs, such as zinc oxide and titanium dioxide, have been used to control pests through their photocatalytic and insecticidal properties. Carbon-based NPs, such as graphene and carbon nanotubes, have also shown potential in controlling pests through their physical and chemical properties (Sabry et al., 2018).

4. **Nanosensors:** They are a unique category of nanopesticides that function as early warning systems for pest infestations. They are designed to detect and monitor pests in real-time and trigger the release of active ingredients to control the pests (Sekhon, 2014). Nanosensors consist of a sensing element, a transducer, and an actuator that releases the active ingredient. Nanosensors have shown promising results in detecting and controlling a wide range of pests, including aphids, thrips, and spider mites (Table 5.3) (Sabry et al., 2018).

TABLE 5.3 Nanopesticides and Their Formulations, Size, and Targeted Pests

Nanopesticides	Formulations	size	Target Pests	References
Nanoemulsion	A homogenous mixture of oil/fat and oil-soluble drugs	50–1000 nm	Mosquito, flies	Abd-Elsalam et al., (2018)
Nanocapsules	Polymer-coating with active ingredients	10–1000 nm	Aphids, mites, whiteflies	Letchford and Burt (2007)
NPs	Organic and inorganic small particles	1–100 nm	*D. melanogaster* (larva, pupa, adult)	Sabry et al., (2018)
Nanosensors	Sensing element, transducer and actuator	10–9 m/10,000 mm	Aphids, thrips, spider, mites	Sabry et al., (2018); Sekhon (2014)

5.2.4 *ON THE BASIS OF PROPERTIES*

1. **Nanopesticide Sensitive to Light:** Light-sensitive nano insecticides are one form of pesticide. The pesticide may be delivered in a targeted and controlled manner by using these nanostructures or

NPs, which can be designed to release it in reaction to particular light wavelengths. Light-sensitive compounds like luminophore are treated with insecticides before being exposed to light. The atomic and molecular Pesticides break down and are released as a result of the altered characteristics of these carrier materials. Light-sensitive insecticides can be created by certain light-sensitive compounds, which can turn light into heat energy. Coumarin is one of the carriers for light-sensitive materials (Djiwanti and Kaushik, 2019). Fipronil was coated on coumarin nanocarrier, causing the pesticide to release in response to light. Fipronil, a nano pesticide, successfully controls aedes larva when it is exposed to blue light and sunshine (Anandhi et al., 2020).

2. **Nanopesticide Sensitive to pH:** There are two types of pH-sensitive nanopesticides, where pesticides are coated on a carrier substance. The first type, monoacid, becomes ionized or deionized in the pH range of 4–8, releasing insecticide at neutral and base pH while receiving protons at low pH. The second type has a base pH-responsive polymer with an amino group that takes protons in an acidic environment and releases them subsequently. Researchers obtained a pH-responsive alginate nanocarrier for cypermethrin by interacting alginate polymer with calcium ions, which promotes their cross-linkage in an acidic pH. This cross-linkage makes the inside of NPs more hydrophobic and slows down the release rate of insecticide (Patel et al., 2018).

   - **Enzyme Sensitive Pesticide:** Changes in plant enzymes are just one of the alterations that take place when a pest interacts with plants. When stimulated, enzyme sensitive polymers can respond chemically and physically. The polymer capsule components shatter under the action of cutinize and liberate pesticide when Lepidopteran insects attack the crop as a result of a change in plant enzymes (Guo et al., 2015). Enzyme-sensitive nanopesticides can be intended to respond to a broad range of enzymes, counting those produced by the pests themselves or by other organisms in the environment. When the NPs encounter the enzyme, they undergo a chemical reaction that triggers the release of the pesticide (Anandhi et al., 2020).

- **Temperature Sensitive Nano Pesticide:** Avermectin insecticide was created by Liang et al., (2018) using silica-coated NPs that have chitosan that responds to temperature. Raising the temperature resulted in a greater production of avermectin, with a discharge of 18.85% at 250°C and 34.21% at 500°C (Anandhi et al., 2020).

Rychter (2019) formulated chitosan/glyphosate (CH/GL) nanopesticides, and the findings indicate that by adjusting the ratio of CH/GL in the formulation, as well as the molecular weight and viscosity of chitosan, it is possible to regulate the release rate of glyphosate, phytotoxicity, and herbicidal effectiveness, thereby creating a safer herbicide.

5.3 CONCLUSION AND THE FUTURE PERSPECTIVE

Nanopesticides have the potential to transform modern agriculture by providing a safer and more effective alternative to traditional pesticides. The four categories of nanopesticides discussed in this chapter—nanoemulsions, nanocapsules, nanoparticles (NPs), and nanosensors—offer a wide range of options for controlling pests and diseases. While challenges remain, such as ensuring the safety and environmental impact of nanopesticides, continued research and development in nanopesticide technology hold great promise for the future of regenerative agriculture.

KEYWORDS

- **nanocapsules**
- **nanoemulsions**
- **nanoformulations**
- **nanoinsecticides**
- **nanoscale**
- **nanotechnology**
- **pesticidal**

REFERENCES

Abd-Elsalam, K. A., & Prasad, R. (2018). *Nanobiotechnology Applications in Plant Protection*. Springer.

Anandhi, S., Saminathan, V. R., Yasotha, P., Saravanan, P. T., & Rajanbabu, V. (2020). Nanopesticides in pest management. *Journal of Entomology and Zoology Studies, 8*, 685–690.

Djiwanti, S. R., & Kaushik, S. (2019). Nanopesticide: Future application of nanomaterials in plant protection. In *Plant Nanobionics: Volume 2, Approaches in NPs, Biosynthesis, and Toxicity* (pp. 255–298).

Giongo, A. M. M., Vendramim, J. D., & Forim, M. R. (2016). Evaluation of neem-based nanoformulations as alternatives to control fall armyworm. *Ciência e Agrotecnologia, 40*, 26–36.

Guan, H., Chi, D., Yu, J., & Li, H. (2010). Dynamics of residues from a novel nano-imidacloprid formulation in soybean fields. *Crop Protection, 29*(9), 942–946.

Guo, M., Zhang, W., Ding, G., Guo, D., Zhu, J., Wang, B., & Cao, Y. (2015). Preparation and characterization of enzyme-responsive Emamectin benzoate microcapsules based on a copolymer matrix of silica–epichlorohydrin–carboxymethylcellulose. *RSC Advances, 5*(113), 93170–93179.

Hossain, F., Follett, P., Salmieri, S., Vu, K. D., Harich, M., & Lacroix, M. (2019). Synergistic effects of nanocomposite films containing essential oil nanoemulsions in combination with ionizing radiation for control of rice weevil *Sitophilus oryzae* in stored grains. *Journal of Food Science, 84*(6), 1439–1446.

Huang, B., Chen, F., Shen, Y., Qian, K., Wang, Y., Sun, C., & Cui, H. (2018). Advances in targeted pesticides with environmentally responsive controlled release by nanotechnology. *Nanomaterials, 8*(2), 102.

Jampílek, J., Kráľová, K., Campos, E. V., & Fraceto, L. F. (2019). Bio-based nanoemulsion formulations applicable in agriculture, medicine, and the food industry. In *Nanobiotechnology in Bioformulations* (pp. 33–84).

Kah, M., Beulke, S., Tiede, K., & Hofmann, T. (2013). Nanopesticides: State of knowledge, environmental fate, and exposure modeling. *Critical Reviews in Environmental Science and Technology*.

Katagi, T. (2012). Environmental behavior of synthetic pyrethroids. In *Pyrethroids: From Chrysanthemum to Modern Industrial Insecticide* (pp. 167–202).

Lade, B. D., Gogle, D. P., Lade, D. B., Moon, G. M., Nandeshwar, S. B., & Kumbhare, S. D. (2019). Nanobiopesticide formulations: Application strategies today and future perspectives. In *Nano-Biopesticides: Today and Future Perspectives*. (pp. 179–206). Academic Press.

Letchford, K., & Burt, H. (2007). A review of the formation and classification of amphiphilic block copolymer nanoparticulate structures: Micelles, nanospheres, nanocapsules, and polymersomes. *European Journal of Pharmaceutics and Biopharmaceutics, 65*(3), 259–269.

Liang, W., Yu, A., Wang, G., Zheng, F., Hu, P., Jia, J., & Xu, H. (2018). A novel water-based chitosan-La pesticide nanocarrier enhancing defense responses in rice (*Oryza sativa* L) growth. *Carbohydrate Polymers, 199*, 437–444.

Pandey, S., Giri, K., Kumar, R., Mishra, G., & Raja Rishi, R. (2018). Nanopesticides: Opportunities in crop protection and associated environmental risks. *Proceedings of the National Academy of Sciences, India Section B: Biological Sciences, 88*, 1287–1308.

Patel, S., Bajpai, J., Saini, R., Bajpai, A. K., & Acharya, S. (2018). Sustained release of pesticide (Cypermethrin) from nanocarriers: An effective technique for environmental and crop protection. *Process Safety and Environmental Protection, 117*, 315–325.

Rychter, P. (2019). Chitosan/glyphosate formulation as a potential, environmentally friendly herbicide with prolonged activity. *Journal of Environmental Science and Health, Part B, 54*(8), 681–692.

Sabry, A. K. H., & Ragaei, M. (2018). Nanotechnology and their applications in insect pest control. In *Nanobiotechnology Applications in Plant Protection* (pp. 1–28).

Salma, U., Chen, N., Richter, D. L., Filson, P. B., Dawson-Andoh, B., Matuana, L., & Heiden, P. (2010). Amphiphilic core/shell NPs to reduce biocide leaching from treated wood: 1–leaching and biological efficacy. *Macromolecular Materials and Engineering, 295*(5), 442–450.

Sekhon, B. S. (2014). Nanotechnology in agri-food production: An overview. In *Nanotechnology, Science and Applications* (pp. 31–53).

Song, S., Liu, X., Jiang, J., Qian, Y., Zhang, N., & Wu, Q. (2009). Stability of triazophos in a self-nanoemulsifying pesticide delivery system. *Colloids and Surfaces A: Physicochemical and Engineering Aspects, 350*(1–3), 57–62.

Taylor, T. M., Weiss, J., Davidson, P. M., & Bruce, B. D. (2005). Liposomal nanocapsules in food science and agriculture. *Critical Reviews in Food Science and Nutrition, 45*(7, 8), 587–605.

Wang, D., Saleh, N. B., Byro, A., Zepp, R., Sahle-Demessie, E., Luxton, T. P., & Su, C. (2022). Nano-enabled pesticides for sustainable agriculture and global food security. *Nature Nanotechnology, 17*(4), 347–360.

Zhu, L., Chen, L., Gu, J., Ma, H., & Wu, H. (2022). Carbon-based nanomaterials for sustainable agriculture: Their application as light converters, nanosensors, and delivery tools. *Plants, 11*(4), 511.

CHAPTER 6

A Systematic Approach to Phytoremediation: Using Nanocarriers and Its Current Perspectives

VINAY ASERI,[1] RUSHIKESH L. CHOPADE,[1] APOORVA SINGH,[1] KUMUD KANT AWASTHI,[2] MAHIPAL SINGH SANKHLA,[2] and GARIMA AWASTHI[2]

[1]*Department of Forensic Science, Vivekananda Global University, Jaipur, Rajasthan, India*

[2]*K. R. Mangalam University, Gurugram, Haryana, India*

ABSTRACT

The quality of the land, water, and atmosphere are all impacted by environmental contamination. Considerable effort have lately been made to reduce pollution-causing sources and repair or improve natural soil and water resources. Nano-phytoremediation is more efficient and less hazardous than chemical and physical approaches. They are becoming more and more significant in this field as well as in academia. Different plant species have demonstrated the ability of nanoparticles to remediate soil and water. It is also to be expected that new environmental cleanup methods will be significantly influenced by the most recent biotechnology research. The aim of this chapter is to immediately summarize the most significant advancements in phytoremediation. Plants have considerable development potential in perspective of their phytoremediation qualities. These plants take in contaminants from the soil and water as well as other natural resources. New strategies

Nanopesticides: A Perspective from Welfare to Risk Assessment.
Garima Awasthi, Payal Lodha, and Gajendra Pal Singh (Eds.)

DOI: 10.1201/9781003605874-6

that are targeted by pollutants include segmenting enzyme activity into plant structures. The majority of the contaminants, including heavy metals, are covered by industrial effluents and have been the subject of epidemiological research. It is fascinating to research the risks of these pollutants in order to develop prevention action against these pollutants. This will help us understand the fate of nanomaterials in the environment and environmental sustainable techniques to remove them from the atmosphere.

6.1 INTRODUCTION

Soil contamination involves the pollution of soil by various anthropogenic activities, including agriculture, industry, and wastewater, which add nutrients, pesticides, and particulates to the soil. Additionally, urbanization and industrial processes further contaminate soil with solid wastes, heavy metals, solvents, and other persistent organic and inorganic substances (Raj et al., 2021; Akmal et al., 2022; Ghani et al., 2022). Contamination arises from both point and non-point sources, including soluble salts (natural and synthetic), pesticides, disposal of urban and industrial waste in agricultural areas, and excessive use of fertilizers. Phytoremediation is defined as the use of green plants, associated soil microorganisms, soil renewal techniques, and agronomic practices to degrade, contain, or transform hazardous chemicals into harmless compounds (Alves et al., 2022; Chaurasia et al., 2023; Bala et al., 2022). This approach enables *in situ* remediation of contaminants from soils, sediment buildup, surface water, and groundwater, offering a cost-effective, solar-driven cleanup method (Gulzar and Mazumder, 2022; Mohanty et al., 2022; Singh et al., 2023). Phytoremediation combines plants' capacity to absorb contaminants into their tissues with climate-adaptive strategies (Rajput et al., 2022). Plants possess an inherent ability to accumulate, degrade, or concentrate chemicals from soil, water, and air, making the removal of toxic metals from the environment a primary goal of phytoremediation. Ongoing research seeks to refine the physiological and molecular mechanisms of phytoremediation for more effective cleanup. Addressing food waste through phytoremediation can also contribute to sustainable waste management.

Point sources of contamination include radiation, industrial effluents, and solid emissions from projects, vehicle usage, and metal pollutants from refining and mining. Contaminants from these sources can be dispersed through air, soil, and water systems, eventually entering ecosystems where they negatively impact human health directly or indirectly (Gulzar and Mazumder, 2022). While metals like copper, iron, manganese, and zinc are essential for biological processes, excess concentrations can lead to toxicity, as seen with metals like cadmium, nickel, and mercury, which have no physiological role and can cause harmful disorders. The ingestion of high-risk metals, such as cadmium, chromium, mercury, and lead, poses significant health risks, including kidney and lung damage, bone deterioration, and liver dysfunction (Gorovtsov et al., 2021; Rajput et al., 2022; Desiante et al., 2021). Although soil acts as a natural filtration system through adsorption, precipitation, and biological decomposition, it has limited capacity to absorb and process these contaminants (Filote et al., 2021). Unlike air pollution, which has direct impacts on human health, soil pollution indirectly affects humans and other animals by contaminating the food chain.

Phytoremediation offers a promising solution for removing various contaminants in situ. Techniques like rhizoremediation and microbe-assisted phytoremediation have shown effectiveness in degrading common soil pollutants, especially when paired with suitable agronomic practices and improvement methods. Phytoremediation is considered an environmentally responsible alternative to the conventional physical and chemical treatment methods that are often harmful to ecosystems (Chiapusio et al., 2007; Nguyen et al., 2022). Advances in phytoremediation now include creating genetically modified plants capable of removing specific chemicals. Phytoremediation uses plants in direct or indirect ways to extract or neutralize pollutants in soil and water. Plants and rhizosphere bacteria can target and transform harmful chemicals in soils, sediments, groundwater, surface water, and even the atmosphere. Phytoremediation is thus a practical, non-invasive, and cost-effective method for soil cleanup.

Nanotechnology, a broad area of study, focuses on materials and applications at very small scales, typically less than 100 nanometers (nm). Nanoparticles exhibit unique properties compared to bulk materials due to their small size, which enables electron concentration and quantum effects. These properties have attracted attention across multiple fields. For example, nanoparticles serve as efficient gas adsorbents, fillers, and carriers for various inorganic chemicals used in environmental remediation.

Nanoparticles can be categorized as either organic or inorganic. Inorganic nanoparticles include metal nanoparticles (e.g., gold and silver), magnetic nanoparticles, and semiconductor nanoparticles, while organic nanoparticles may include carbon nanoparticles (e.g., fullerenes) or those derived from decomposed natural materials like orange peels or potato skins. The use of nanoparticles in environmental applications is growing due to their large surface area relative to bulk materials, which enhances their reactivity and suitability for targeted environmental responses (Romeh, 2022; Sarma et al., 2022).

6.2 TECHNOLOGICAL ADVANCES IN PHYTOREMEDIATION

Nano bioremediation is the process of eliminating conventional toxins (such heavy metals) from contaminated areas using nanoparticles/nanomaterial framed by plant, natural components, and Nano bioremediation (NBR). NBR is a newly developed framework for removing dangerous compounds in order to facilitate natural cleansing. Burning, bioremediation, substance remediation, and authentic remediation are the current stages of advancement for cleaning up contaminated complaints. With ongoing improvements, bioremediation now provides a financially viable and environmentally safe option to remove pollutants from the atmosphere (Ojuederie et al., 2022; Song et al., 2022).

6.3 NANOPARTICLE INTERACTIONS WITH PLANTS

The size, type, and design of nanoparticles strongly influence their rate of absorption by plants. Primarily, the size of the nanoparticle is a critical factor in determining its ability to penetrate plant tissues and spread from roots to other parts of the plant. Additionally, the entry mechanism of nanomaterials affects their absorption efficiency by plants (Oladoye et al., 2022; Gavrilescu, 2022). After entering the roots, nanoparticles can migrate within plant tissues through two main pathways: Apoplastic transport, which occurs outside of the plasma membrane and along xylem vessels, and symplastic transport, which involves the movement of water and materials through the cytoplasm and sieve plates. Environmental conditions also significantly impact nanoparticle uptake. For instance, temperature affects root growth and nutrient absorption, with notable differences observed between root structures in field-grown plants versus those in nursery environments. Developing a novel hyperaccumulator—plant species that can store high concentrations of specific contaminants—is crucial for advancing phytoremediation techniques, specifically in the area of phytoextraction (Singh and Khan, 2022).

6.4 NANO-PHYTOREMEDIATION

Nano-phytoremediation combines phytoremediation and nanotechnology to improve the environment. Nanotechnology increases the efficacy of phytoremediation, and nanoparticles can also be used to clean up soil and water that have been hazardous metal, organic, and inorganic chemical contamination. Recent studies have shown that ordinary new chemicals like atrazine, molinate, and chlorpyrifos may be contaminated by zerovalent irons as tiny as a nanometer. In protein-based bioremediation, phytoremediation can also be mixed with nanoparticles (Song et al., 2022). The use of nanoparticles to clean up the environment is developing swiftly. Iron nanoparticles are widely used throughout the cycle of clean-up. Some researcher looks at whether organic contaminants (PCBs, PAHs, PPCPs, and organic solvents) in the soil may be decontaminated by nanomaterials. A combination of nanotechnology and phytotechnology for TNT remediation from soil is more efficient than either technology individually, claim soil clean of nanoparticles (Prakash et al., 2023).

6.4.1 FACTORS AFFECTING CONTAMINANT NANO-PHYTOREMEDIATION

Several factors can influence how a drug is absorbed and distributed by living plants. These include the physical and chemical characteristics of the mixtures (e.g., water solubility, vapor pressure, and atomic weight); environmental conditions (e.g., temperature, pH, organic matter, and soil moisture content); and aspects of the plant itself (e.g., type of root development and nature of compound interactions). In Figure 6.1, a schematic outline of the factors affecting nano-phytoremediation is shown.

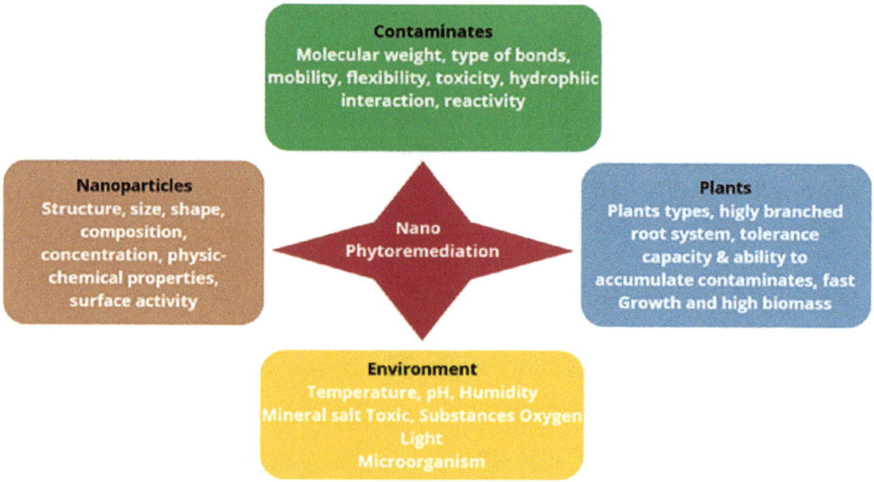

FIGURE 6.1 Schematic representation of phytotechnology.

6.4.2 PLANTS WITH THE BEST PROPERTIES FOR NANO-PHYTOREMEDIATION

Ideal plants for nano-phytoremediation exhibit rapid growth and high biomass production (enhanced development and productivity); an extensive root surface area with well-developed lateral roots (broad, highly branched root systems); and the ability to tolerate or accumulate contaminants (Gulzar and Mazumder, 2022) (high resistance to toxicity). They should also have a high threshold for the hyperaccumulation of foreign compounds (inorganic, organic, and heavy metals), particularly in above-ground tissues (sink potential). Additionally, they should be easy to harvest (considering the utility of the sink organ) and not suitable for human or animal consumption.

6.5 CHOOSING THE RIGHT NANOPARTICLES FOR PHYTOREMEDIATION

The following characteristics are desirable for nanoparticles used in phytoremediation: they should not be hazardous to plants; they should promote increased biomass, root-to-shoot ratio, plant height, and seedling development; and they should significantly enhance the production of phytoenzymes. Additionally, they should be able to improve the efficacy of plant growth chemicals and possess the capacity to bind contaminants, thereby increasing their bioavailability to plants.

The remediation of degraded soils can benefit from a combined nano- and phytoremediation approach (phyto-nano-treatment). In order to clean up contaminated areas, nano-phytoremediation technologies (plant- and nanoparticle-based approaches) employ either naturally occurring or genetically engineered plants (Daraz et al., 2022). The use of carefully selected nanoparticles has been shown to result in improved plant growth, while nano-enhancement boosts phytoremediation capacity, enabling substantial pollutant removal from the soil environment (Sharma et al., 2022).

Many nanoparticles have also been recognized as plant growth promoters because they can enhance the production of plant hormones and improve pollutant uptake. Furthermore, it has been observed that several plant species respond particularly well to nano-treatment methods. Numerous studies have consistently highlighted the positive effects of metal and metal oxide nanoparticles on the growth of higher plants.

Despite the potential advantages of nanocarbon, a major challenge to its application in soil remediation is its potential to harm plants and other soil organisms (Ganguly et al., 2023). The application of nanomaterials is site-specific, and their effectiveness is influenced by factors such as the type and concentration of the contaminant, the size and charge of the nanoparticle, the nature of the compound, and various physical and chemical soil properties, including soil aggregates and texture in relation to the size of the nanomaterial.

Different techniques are employed for applying nanomaterials to soil during the remediation process (Su et al., 2022). These include hydrodynamic dispersion, pressure pulse generation, distribution, and pressurized injection. Figure 6.2 provides a schematic outline of the application of nanoparticles in phytoremediation.

6.6 POLLUTANTS IN SOIL REMEDIATED BY NANO-PHYTOTECHNOLOGY

The combined use of nanoparticles and phytoremediation systems is essential for advancing nano-phytoremediation in the remediation of polluted soils. The integration of phytoremediation with nanoparticles represents a significant step toward effective soil sanitization (Rajput et al., 2022). Numerous studies have demonstrated that various nanoparticles and nanomaterials can detoxify or repair soil contamination by common, inorganic, and heavy metal pollutants. Bimetallic nanoparticles (Pd/Fe), magnetite nanoparticles (nFe_3O_4), and nano zerovalent iron (nZVI) are especially effective at rapidly degrading conventional pollutants, including chlorpyrifos, atrazine, lindane, pentachlorophenol, trichloroethylene (TCE), pyrene, polychlorinated biphenyls (PCBs), and 2,4-dinitrotoluene. Nano titanium dioxide ($nTiO_2$) reduces the half-lives of pyrene and phenanthrene in soil and can degrade hazardous chemicals (e.g., diuron, p,p′-DDT, pyrene, and phenanthrene) via photocatalysis, redox reactions, and thermal degradation when illuminated (He et al., 2016). Additionally, persistent contamination by atrazine has been addressed through PEI-copper nanoparticles retained on montmorillonite and sand (Ma and Wang, 2010).

Nanoparticles can have mixed effects on soil microorganisms, influencing plants both positively and negatively. For instance, the application of silver nanoparticles (Ag) at concentrations of 0.1–10 mg kg^{-1} produced an unfavorable outcome, whereas positive responses in soil microorganisms were sometimes observed with iron oxide (FeO) nanoparticles at 1 or 10 mg kg^{-1} (Pillai and Kottekottil, 2016). Nano-phytoremediation can significantly reduce soil contamination and has shown strong potential for remediating heavy metal pollution through enhanced plant uptake in atypical soil mixes. Nanomaterial-aided phytoremediation is also highly effective at removing conventional pollutants and repairing contaminated sites. Recent studies indicate that nanoparticles can enhance plants' ability to tolerate stress in both *ex situ* and *in situ* conditions, thus improving their phytoremediation potential (Souri et al., 2017).

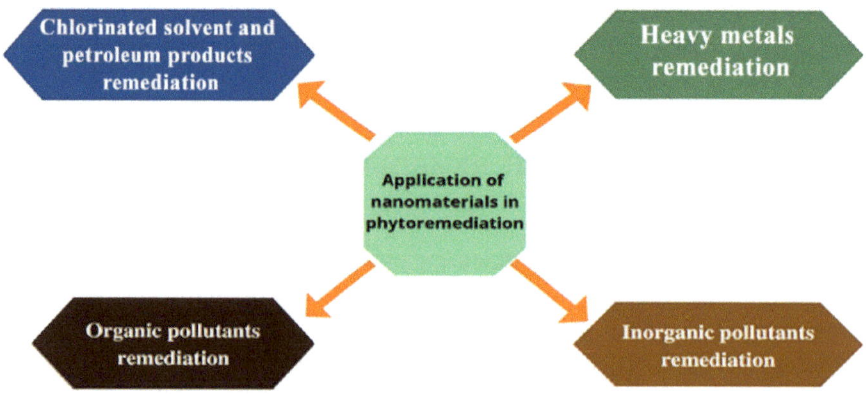

FIGURE 6.2 Application of nanomaterials in phytoremediation.

6.7 PLANT-ASSISTED REMEDIATION IN CONTAMINATED SOIL USING NANOMATERIAL

It depends on the employment of nanoparticles and phytoremediation techniques (such as phytochange, phytoextraction, and phytodegradation) using plants which plant species will be used in the nano-phyto mechanical remediation (nano-phytoremediation) framework for cleaning contaminated soil. This innovation has demonstrated a remarkable capability for contaminant accumulation and biological remediation of common pollutants. Plant species accelerate the release of toxins from contaminated soil, which requires more effort than usual. Previous studies have shown that the combination of plant species and nanoparticles caused significant new chemicals to be released from contaminated soil (Burken, 2001; Schwab et al., 2016).

In order to evaluate the effectiveness of new chemicals for nano-phytoremediation, various plant species and groups of nanoparticles are used. Their suitability for release is shown in Table 6.1. Nanoparticles interact with plants, causing many physiological and morphological changes. The response of the plants will depend on the type, size, and speciation of the nanoparticles, as well as the specific plant species used (Schroeder et al., 2009). Their physical-chemical composition, size, shape, surface coating, reactivity, and especially the substrate on which they are applied—which varies from one plant to another—determine their suitability, though not permanently.

Growing evidence suggests that different nanoparticles enhance plant development and growth in various studies. In particular, silver nanoparticles increased the production of ABA and GA phytohormones in plants by over 90%, which helps plants to endure stress and enhances the uptake of nutrients and water for improved growth. Similarly, salicylic-acid-based nanoparticles promote plant growth and improve the phytoremediation ability of plants under arsenic (As) stress (Satapanajaru et al., 2008). After applying MnNP to the soil, Pradhan et al. observed beneficial effects on root and shoot expansion, as well as nitrogen uptake, metabolism, and plant development (Singh et al., 2012). Incorporating nano-fertilizers into nanoparticles in this way helps to increase soil nutrient availability, accelerate soil productivity, and promote enzymatic activity in plants, enhancing chlorophyll content and seed germination.

TABLE 6.1 List of Plant Species and Nanoparticles Used in a Nano Phytoremediation System to Remove Pollutants

Plant species	Used NPs	Contaminants	Removal efficiency	References
Alpinia calcarata	nZVIs	Endosulfan	≈100 (%)	Li et al., (2016); Makarova et al., (2000)
Ocimum sanctum	nZVIs	Endosulfan	76.28 ± 0.19 (%)	Gulzar and Mazumder, (2022)
Cymbopogon citratus	nZVIs	Endosulfan	86.16 ± 0.09 (%)	Song et al., (2022)
Populus deltoides	Fullerene (nC60)	Trichloroethylene	82 (%)	Kalidhasan et al., (2017)
Isatiscappadocica	SANPs	Arsenic	705 ppm and 1188 ppm As accumulate in roots and shoots, respectively	Kalidhasan et al., (2017)

6.8 CHALLENGES OF NANO-PHYTOREMEDIATION

Studies on phytoremediation aided by nanomaterials are incredibly scant. Since only microcosm experiments have been conducted so far, future research must incorporate more sensible assessments, and improved field knowledge is crucially needed. Long-term testing is anticipated to reveal the real effects of nanoparticles on the phytoremediation cycle and the status of the soil's richness. Nanomaterials can increase this and potentially decrease their transportability; thus polymers or other coatings are intended to modify their bioavailability or adaptability. Analyzing the effects and success of applying nanomaterials to contaminated soil is necessary. Appropriately environmentally stable nanomaterial should be considered, since viable nano-phytoremediation largely depends on meteorological circumstances. Nanophytoremediation is a valid method for problems with low levels of pollution due to specious plant growth in extremely polluted soils. Growing agromining, which may be used to remove toxins even from harvestable plant biomass, will be supported by a general understanding of the poisons absorbed by plants from soil (Table 6.2).

6.9 CONCLUSION

Nanophytoremediation technology is a relatively new field within environmental remediation processes. Most studies have been confined to laboratory and pot culture settings, with few conducted at full scale. Differences between lab/pot cultivation experiments and field experiments

TABLE 6.2 A List of Possible Nanoparticles in the Process of Phytoremediation

Sl. No.	Nanoparticles	Plant Species	Beneficiary Concentration(s)	Effects	References
1.	MnNPs	*Vigna radiata*	50–1000 mg L^{-1}	Expanded nitrogen take-up, root and shoot prolongation	Roberts (2003)
2.	MWCNTs	*Lycopersicon esculentum*	10–40 mg L^{-1}	Upgraded the seed germination and development	Burken and Schnoor (1996)
3.	AgNPs	*Zea mays* L.		Expanded creation of ABA and GA phyto-chemicals	Ahmadpour et al., (2012)
4.	AgNPs	*Boswellia ovalifoliolata*	10–30 g mL^{-1}	Expanded germination and seedling development	Sajid et al., (2015)
5.	SANPs	*Isatis cappadocica*		Improve growth	Sajid et al., (2015); Makarova et al., (2000)
6.	Aluminum oxide NPs	*Arabidopsis thaliana*	400-4000 mg L^{-1}	Root length	Makarova et al., (2000)
7.	CO_3O_4 NPs	*Raphanus sativus* L.	5 g L^{-1}	Root elongation	Kalidhasan et al., (2017)
8.	Fe_2O_3 NPs	*Arachis hypogaea*	2–1000 mg kg^{-1}	Expanded root length, plant level, biomass.	Kalidhasan, et al., (2017)
9.	MnOx NPs	*Lactuca sativa*	10 mg L^{-1}	Worked on the development of lettuce seedlings by upgrading root stretching	Khan and Bano (2016)
10.	SWCNTs	*Allium cepa, Cucumis sativus*	9, 56, 315, and 1750 mg L^{-1}	Root elongation	Savithramma et al., (2012)
11.	Graphene oxide	*Vicia faba* L.	400 and 800 mg L^{-1}	Germination	Savithramma et al., (2012)
12.	ZnO NPs	*Arachis hypogaea*	1000 ppm	Germination, stem, root development, and yield.	Bao-Shan et al., (2004)
13.	GNPs	*Arabidopsis thaliana*	10 and 80 µg mL^{-1}	Germination, shoot and root development.	Dimkpa et al., (2012)
14.	TiO2 NPs Root	*Arabidopsis thaliana*	400 mg L^{-1}	Length	Dimkpa et al., (2012)
15.	SiO2 NPs	*Changbai larch*	500 µL L^{-1}	Expanded shoot level, root collar breadth, root length	Awasthi et al., (2022); Raveendran et al., (2022)
16.	CuO NPs	Common wheat	500 mg kg^{-1}	Expanded biomass	Nagar et al., (2022)

were observed due to variables like temperature, pH, soil type, nutrients, and soil moisture content, all of which influence outcomes in real field environments. This review highlights recent discoveries and research on using nanomaterials to accelerate the phytoremediation of polluted soil. Nanomaterials have demonstrated potential in removing pollutants, promoting plant growth, and increasing the phytoavailability of contaminants, thereby facilitating the phytoremediation of polluted soils.

6.10 FUTURE WORK

Nanotechnology offers green and eco-friendly options for environmental cleanup and management without causing harm to the natural environment. Various plants, organisms, and life forms, known as "hyperaccumulators," have been identified for their high capacity to absorb large quantities of metals. These hyperaccumulators include certain plants, fungi, and microbial species, which show promise for bioremediation of metal-contaminated areas. Nanomaterials can be leveraged in various ways to enhance the release and uptake of pollutants by these species. To fully understand the impact of nanomaterials on plants and the environment, it is essential to investigate the mechanisms by which they interact with the ecosystem. Advancing agronomic practices for effective remediation will require the careful selection of suitable plant species and nanomaterials to target specific contaminants. As a field with vast potential, nanophytoremediation stands at the intersection of multiple disciplines and offers opportunities for innovative environmental solutions. This cross-disciplinary approach holds promise for the positive impact of nanotechnology in addressing environmental contamination and advancing sustainable practices.

KEYWORDS

- **agribusiness**
- **agromining**
- **bioremediation**
- **nanoparticles**
- **nano-phytoremediation**
- **nanotechnology**

REFERENCES

Ahmadpour, P., Ahmadpour, F., Mahmud, T. M. M., Abdu, A., Soleimani, M., & Hosseini Tayefeh, F. (2012). Phytoremediation of heavy metals: A green technology. *Environmental Engineering and Management Journal, 11,* 14036–14043.

Akmal, M., Younus, A., Wakeel, A., Jamil, Y., & Rashid, M. A. (2022). Synthesis and application of optimized ZnO nanoparticles for improving yield and Zn content of rice (*Oryza sativa* L.) grain. *Science Progress,* 1–14.

Alves, A. R. A., Yin, Q., Oliveira, R. S., Silva, E. F., & Novo, L. A. B. (2022). Plant growth-promoting bacteria in phytoremediation of metal-polluted soils: Current knowledge and future directions. *Environmental Biotechnology,* 156435.

Awasthi, G., Nagar, V., Mandzhieva, S., Minkina, T., Sankhla, M. S., Pandit, P. P., Aseri, V., Awasthi, K. K., Rajput, V. D., & Bauer, T. (2022). Sustainable amelioration of heavy metals in soil ecosystem: Existing developments to emerging trends. *Environmental Monitoring and Assessment, 12,* 85.

Bala, S., Garg, D., Thirumalesh, B. V., Sharma, M., Sridhar, K., Inbaraj, B. S., & Tripathi, M. (2022). Recent strategies for bioremediation of emerging pollutants: A review for a green and sustainable environment. *Toxics, 10,* 484.

Bao-Shan, L., Diao, S. Q., Li, C. H., Fang, L. J., Qiao, S. C., & Yu, M. (2004). Effect of TMS (nanostructured silicon dioxide) on growth of Changbai larch seedlings. *Forest Research, 15,* 138–140.

Burken, J. G. (2001). Advancement of phytoremediation. *Practice Periodical of Hazardous, Toxic, and Radioactive Waste Management, 5,* 120–121.

Burken, J. G., & Schnoor, J. L. (1996). Phytoremediation: Plant uptake of atrazine and role of root exudates. *Environmental Science & Technology, 122,* 958–963.

Chaplygin, V. A., Minkina, T. M., Mandzhieva, S. S., Nazarenko, O. G., Zimulina, I. V., Bauer, T. V., Litvinov, Y. A., & Rajput, V. D. (2021). Heavy metals in agricultural crops of Rostov region through the example of soft wheat (*Triticum aestivum*).

Chaurasia, P. K., Sharma, N., Kumari, S., Yadav, M., Singh, S., Mani, A., Yadava, S., & Bharati, S. L. (2023). Fungal assisted bio-treatment of environmental pollutants with comprehensive emphasis on noxious heavy metals: Recent updates. *Environmental Science and Pollution Research, 120,* 57–81.

Chiapusio, G., Pujol, S., Toussaint, M. L., Badot, P. M., & Binet, P. (2007). Phenanthrene toxicity and dissipation in rhizosphere of grassland plants (*Lolium perenne* L. and *Trifolium pratense* L.) in three spiked soils. *Chemosphere, 294,* 103–112.

Daraz, U., Li, Y., Ahmad, I., Iqbal, R., & Ditta, A. (2022). Remediation technologies for acid mine drainage: Recent trends and future perspectives. *Journal of Environmental Management,* 137089.

Desiante, W. L., Minas, N. S., & Fenner, K. (2021). Micropollutant biotransformation and bioaccumulation in natural stream biofilms. *Environmental Pollution, 193,* 116846.

Dimkpa, C. O., McLean, J. E., Latta, D. E., Manangón, E., Britt, D. W., Johnson, W. P., Boyanov, M. I., & Anderson, A. J. (2012). CuO and ZnO nanoparticles: Phytotoxicity, metal speciation, and induction of oxidative stress in sand-grown wheat. *Journal of Nanoparticle Research, 14,* 1–15.

Filote, C., Roşca, M., Hlihor, R. M., Cozma, P., Simion, I. M., Apostol, M., & Gavrilescu, M. (2021). Sustainable application of biosorption and bioaccumulation of persistent pollutants in wastewater treatment: Current practice. *Processes, 9,* 1696.

Ganguly, R., Mathur, A., & Singh, R. P. (2023). Bioremediation of heavy metals from ecosystem nanotechnological perspectives. In *Bioremediation of Toxic Metal (loid)s* (CRC Press).

Gavrilescu, M. (2022). Enhancing phytoremediation of soils polluted with heavy metals. *Current Opinion in Biotechnology, 74,* 21–31.

Ghani, M. I., Saleem, S., Rather, S. A., Rehmani, M. S., Alamri, S., Rajput, V. D., Kalaji, H. M., Saleem, N., Sial, T. A., & Liu, M. (2022). Foliar application of zinc oxide nanoparticles: An effective strategy to mitigate drought stress in cucumber seedlings by modulating antioxidant defense system and osmolyte accumulation. *Chemical Engineering Journal, 289,* 133202.

Gorovtsov, A., Demin, K., Sushkova, S., Minkina, T., Grigoryeva, T., Dudnikova, T., Barbashev, A., Semenkov, I., Romanova, V., & Laikov, A. (2021). The effect of combined pollution by PAHs and heavy metals on the topsoil microbial communities of Spolic Technosols of Lake Atamanskoe, Southern Russia. *Chemosphere,* 1–17.

Gulzar, A. B. M., & Mazumder, P. B. (2022). Helping plants to deal with heavy metal stress: The role of nanotechnology and plant growth promoting rhizobacteria in the process of phytoremediation. *Environmental Science and Pollution Research, 29,* 40319–40341.

He, S., Feng, Y., Ni, J., Sun, Y., Xue, L., Feng, Y., Yu, Y., Lin, X., & Yang, L. (2016). Different responses of soil microbial metabolic activity to silver and iron oxide nanoparticles. *Chemosphere, 147,* 195–202.

Kalidhasan, S., Dror, I., & Berkowitz, B. (2017). Atrazine degradation through PEI-copper nanoparticles deposited onto montmorillonite and sand. *Scientific Reports, 7,* 1415.

Khan, N., & Bano, A. (2016). Modulation of phytoremediation and plant growth by the treatment with PGPR, Ag nanoparticle, and untreated municipal wastewater. *Environmental Science and Pollution Research, 18,* 1258–1269.

Li, Q., Chen, X., Zhuang, J., & Chen, X. (2016). Decontaminating soil organic pollutants with manufactured nanoparticles. *Environmental Science & Technology, 23,* 11533–11548.

Ma, X., & Wang, C. (2010). Fullerene nanoparticles affect the fate and uptake of trichloroethylene in phytoremediation systems. *Environmental Engineering Science, 27,* 989–992.

Makarova, O. V., Rajh, T., Thurnauer, M. C., Martin, A., Kemme, P. A., & Cropek, D. (2000). Surface modification of TiO_2 nanoparticles for photochemical reduction of nitrobenzene. *Journal of Physical Chemistry B, 34,* 4797–4803.

Moameri, M., & Abassi Khalaki, M. (2021). Toxicity/risk assessment of nanomaterials when used in soil treatment. *Nanomaterials for Soil Remediation, 2021*, 87–100.

Mohanty, S., Chakraborty, S., Das, M., & Paul, S. (2022). Role of nanomaterials in phytoremediation of tainted soil. *Phytoremediation Technology for the Removal of Heavy Metals and Other Contaminants from Soil and Water, 2022*, 329–353.

Nagar, V., Singh, T., Tiwari, Y., Aseri, V., Pandit, P. P., Chopade, R. L., Pandey, K., Lodha, P., & Awasthi, G. (2022). ZnO nanoparticles: Exposure, toxicity mechanism, and assessment. *Materialstoday: Proceeding, 69*(Part 1), 56–63.

Nguyen, N. T. T., Nguyen, L. M., Nguyen, T. T. T., Tran, U. P. N., Nguyen, D. T. C., & Tran, T. V. (2022). A critical review on the bio-mediated green synthesis and multiple applications of magnesium oxide nanoparticles. *Nanomaterials, 12,* 137301.

Ojuederie, O. B., Amoo, A. E., Owonubi, S. J., & Ayangbenro, A. S. (2022). Nanoparticles-assisted phytoremediation: Advances and applications. *Environmental Science and Pollution Research, 155,* 155–78.

Oladoye, P. O., Olowe, O. M., & Asemoloye, M. D. (2022). Phytoremediation technology and food security impacts of heavy metal contaminated soils: A review of literature. *Environmental Science and Pollution Research, 288,* 132555.

Pillai, H. P. S., & Kottekottil, J. (2016). Nano-phytotechnological remediation of endosulfan using zero valent iron nanoparticles. *Environmental Technology,* 37, 1–8.

Prakash, O., Maurya, S., Tripathy, P., Sharma, A., Vijay, R., & Pal, S. (2023). Nano-and phytoremediation technique for textile wastewater treatment and successive production of fertilizers. *Metagenomics to Bioremediation: Applications, Cutting Edge Tools, and Future Outlook Developments in Applied Microbiology and Biotechnology, 2023,* 537–559.

Raj, S. N., Anooj, E. S., Rajendran, K., & Vallinayagam, S. (2021). A comprehensive review on regulatory intervention of nanopesticides in agricultural nano formulation and food system. *Food Control, 1239,* 130517.

Rajput, V. D., Minkina, T., Upadhyay, S. K., Kumari, A., Ranjan, A., Mandzhieva, S., Sushkova, S., Singh, R. K., & Verma, K. K. (2022). Nanotechnology in the restoration of polluted soil. *Nanomaterials, 12,* 769.

Rajput, V. D., Yadav, A. N., Jatav, H. S., Singh, S. K., & Minkina, T. (2022). A review on nanobioremediation approaches for restoration of contaminated soil. *Springer Nature.*

Raveendran, V. C., Godara, V., Mistry, D., Sankhla, M. S., Awasthi, K. K., Singh, G. P., Lodha, P., & Awasthi, G. (2022). Nanopesticides: Composition, bioavailability, and release mechanism in the environment. *Toxics, 8,* 116–121.

Roberts, A. (2003). Plasmodesmata and the control of symplastic transport. *Plant Cell & Environment, 26,* 103–124.

Romeh, A. A. (2022). Nano-phytoremediation: Using plants and nanomaterials to environmental pollution remediation. In *Nanotechnology for Environmental Pollution Decontamination* (Apple Academic Press).

Sajid, M., Ilyas, M., Basheer, C., Tariq, M., Daud, M., Baig, N., & Shehzad, F. (2015). Impact of nanoparticles on human and environment: Review of toxicity factors, exposures, control strategies, and future prospects. *Environmental Science and Pollution Research, 22,* 4122–4143.

Sarma, H., Narayan, M., Peralta-Videa, J. R., & Lam, S. S. (2022). Exploring the significance of nanomaterials and organic amendments—Prospect for phytoremediation of contaminated agroecosystems. *Environmental Pollution, 308,* 119601.

Satapanajaru, T., Anurakpongsatorn, P., Pengthamkeerati, P., & Boparai, H. (2008). Remediation of atrazine-contaminated soil and water by nano zerovalent iron. *Journal of Hazardous Materials, 192,* 349–359.

Savithramma, N., Ankanna, S., & Bhumi, G. (2012). Effect of nanoparticles on seed germination and seedling growth of Boswellia ovalifoliolata, an endemic and endangered medicinal tree taxon. *Journal of Medicinal Plants Research, 2,* 2.

Schroeder, W. H., Breuer, U., Stelzer, R., & Gierth, M. (2009). The apoplast and its significance for plant mineral nutrition. *New Phytologist, 149*(2), 167–192.

Schwab, F., Zhai, G., Kern, M., Turner, A., Schnoor, J. L., & Wiesner, M. R. (2016). Barriers, pathways and processes for uptake, translocation and accumulation of nanomaterials in plants–Critical review. *Nanotoxicology, 10,* 257–278.

Singh, G., Bhadange, S., Bhawna, F., Shewale, P., Dahiya, R., Aggarwal, A., Manju, F., & Arya, S. K. (2023). Phytoremediation of radioactive elements, possibilities and challenges: Special focus on agricultural aspects. *Environmental Science and Pollution Research, 25,* 1–8.

Singh, N., & Khan, S. (2022). Fluoride (F) remediation using phytoremediation and nanomaterials. *In Phytoremediation: A Sustainable Approach to Water and Soil Remediation* (pp. 325–348).

Singh, R., Misra, V., Mudiam, M. K. R., Chauhan, L. K. S., & Singh, R. P. (2012). Degradation of γ-HCH spiked soil using stabilized Pd/Fe0 bimetallic nanoparticles: Pathways, kinetics and effect of reaction conditions. *Journal of Hazardous Materials, 237,* 355–364.

Song, U., Kim, B. W., Rim, H., & Bang, J. H. (2022). Phytoremediation of nanoparticle-contaminated soil using the halophyte plant species *Suaeda glauca*. *Environmental Science and Pollution Research, 28,* 102626.

Souri, Z., Karimi, N., Sarmadi, M., & Rostami, E. (2017). Salicylic acid nanoparticles (SANPs) improve growth and phytoremediation efficiency of Isatis cappadocica Desv., under As stress. *Journal of Plant Growth Regulation, 11,* 650–655.

Srivastava, S., Shukla, A., Rajput, V. D., Kumar, K., Minkina, T., Mandzhieva, S., Shmaraeva, A., & Suprasanna, P. (2021). Arsenic remediation through sustainable phytoremediation approaches. *Environmental Technology & Innovation, 11,* 936.

Su, R., Wang, Y., Huang, S., Chen, R., & Wang, J. (2022). Application for ecological restoration of contaminated soil: Phytoremediation. *Journal of Environmental Management,* 100010.

CHAPTER 7

Interaction of Nanopesticides with Plants

APEXA PAREEK, RENU SHARMA, TARA MAHESHWARI, and PAYAL LODHA

Department of Botany, University of Rajasthan, Jaipur, Rajasthan, India

ABSTRACT

In recent years, to meet global sustainable development goals, the use of chemical pesticides has been minimized or banned, placing significant pressure on the agricultural sector to improve food quality and health while meeting the demands of a growing population. Nanopesticides, or nanomaterials—a new approach from nanotechnology—have been introduced to manage various plant diseases, and a growing number of research studies have been published on this topic. To assess the impact of nanopesticides on the environment, whether positive or negative and potentially more reliable than other control strategies, it is essential to understand how nanopesticides interact with plants. In this context, this chapter discusses the mechanisms of nanopesticide absorption, uptake, and translocation in plants, as well as their effects on plant physiology and overall plant responses.

7.1 INTRODUCTION

One of the greatest obstacles globally in utilizing sustainable techniques is the necessity to increase food production to satisfy the hunger of a rising

Nanopesticides: A Perspective from Welfare to Risk Assessment.
Garima Awasthi, Payal Lodha, and Gajendra Pal Singh (Eds.)

DOI: 10.1201/9781003605874-7

population. Population growth continues to rise and has already surpassed 7.37 billion people (World Bank Data, 2015). Among farming communities, the use of chemical fertilizers and pesticides, genetically modified crops, and disease-resistant crop varieties has been popular over the past fifty years as a means of addressing this issue. In reality, the unregulated use of chemical fertilizers, pesticides, and other agrochemical products has significantly increased food production but has also reduced food quality and soil fertility. It poses risks to ecosystem components and to consumers who are directly or indirectly involved in the agrochemical chain (Hayles et al., 2017; Kah et al., 2013; Zamir, 2001). It has been claimed that through leaching, mineralization, and bioconversion, 50–70% of waste remains unused (Bollag et al., 1992). In addition to harming human health, residues from chemical pesticides and fertilizers have destabilized ecosystem sublevels—such as soil microbial flora, parasites, and aquatic environments—due to eutrophication (Conley et al., 2009). Therefore, there is a growing demand to transform traditional agricultural methods into smart practices through the involvement of advanced technologies such as nanotechnology for sustainable agriculture.

After biotechnology, nanotechnology is the truly radical and sustainable innovation for the world. Nanofertilizers, nanopesticides and nanohormones that are collectively referred as nanoagrochemicals are applied form of nanotechnology in agro industry. Nanoscale materials may be great choice for the drafting and development of new innovative nanopesticides and fertilizers. Nanoagrochemicals may improve the quality of agricultural products, reduce the harmful effects of agricultural pesticides on the nature and health of human, boost productivity, and increase food security (Asif et al., 2021; Chaud et al., 2021; Shekhar et al., 2021). Various metals such as Zn, Cu, Ce, Ag, Au, Fe, Si, Ti, and their oxides and non-metals like Carbon, Chitosan and nano-formulation of some chemicals and minerals such phosphorus, urea, sulfur have been transformed into nanopesticides and nanofertilizers (Ghafariyan et al., 2013; Mahajan et al., 2011; Pradhan et al., 2013; Taha et al., 2016).

In last few years, various studies showed that these nanoforms have potential to enhance growth, production, protection against pest and seed germination of various crop by their control and targeting delivery (Delfani et al., 2014; Kashyap et al., 2015; Liu et al., 2006; Parisi et al., 2014). Nano-based agro-chemicals may be viewed as particularly important in terms of potential environmental effect due to their direct and intended use in the environment, since they are the source of diffusion manufactured nanoparticles in the environment (Kah et al., 2013). In this chapter, potential effects of various nanoforms/nanomaterials on growth and development of plants, way of their translocation and interaction with plants will be discussed, so that their impacts on environment will be asset and help in commercialization of various nano forms.

7.2 MECHANISM OF ACTION

7.2.1 UPTAKE OF NANOPARTICLES IN PLANTS

Plants are the main fundamental component as primary producers of an ecosystem and maintain equilibrium of mineral and nutrition through their translocation. Plants get nutrition and minerals from soil, water and environment thus, nanoparticles have many ways or routes such as release via human, direct application, contaminated soil/water, etc., to interact with plants. The reaction between nanoparticles and plants has become inescapable; it is our need to know about the mechanism of uptake and translocation of nanoparticles into plants so that following information could be source for the further development in nanopesticides.

Uptake of nano particles is affected by multiple factors such as size, shape and chemical constitute of nanoparticle itself and affected with plant type, species and site of application (Pérez-de-Luque, 2017; Raliya et al., 2016). Besides all above mention factors, the way nano particles application is very crucial factor that determine effectiveness of up taking nanoparticles by plants: roots absorb water and minerals whereas leaves exchange gases (Schwab et al., 2016). Nanoparticles can be entering into leaves tissues through various methods such as stomata, wounds and direct penetration, etc. (Figure 7.1). The first hindrance in up taking of nano particles is the cell wall of a plant that allows only small size particles (pore size of the cell wall 5–20 nm) to enter plant system. Thus, nanoparticles which have comparable diameter to pore size of the cell wall can creep into the plasma membrane (Moore, 2006; Navarro et al., 2008).

After internalization, nanoparticles can cross the plasma membrane in several ways: endocytosis, where nanoparticles are engulfed by the plasma membrane through invagination and transported to different cellular compartments (Etxeberria et al., 2006); pore formation, where some nanoparticles disrupt the plasma membrane and create new pores

(Wong et al., 2016); and transport through aquaporins, as well as other membrane-embedded carrier proteins and ion channels (Rico et al., 2011). Shen et al. (2010) reported that single-walled carbon nanotubes (SWCNTs) of larger sizes could not penetrate the cell wall of *Arabidopsis thaliana* but formed endocytosis-like structures on the plasma membrane. In contrast, water-soluble SWCNTs could penetrate the cell wall and were internalized through endocytosis in nicotiana plants (Liu et al., 2009). Thus, there is no specific uptake route or mechanism for a particular nanoparticle, as differences in translocation and accumulation mechanisms have been observed in different plant species for the same types of nanoparticles (Zhu et al., 2012).

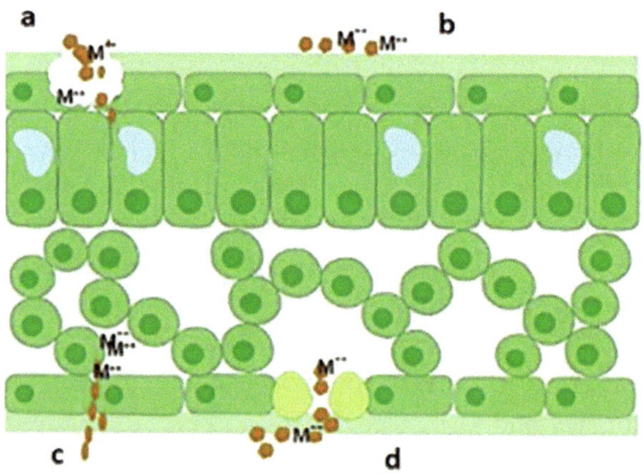

FIGURE 7.1 Various method of entry of nanoparticles into leaves (a) through wounds (b) through cuticle (c) through various cell membrane channels/carriers (d) through stomata.

7.2.2 TRANSLOCATION INTO PLANTS

After entry into cells, nanoparticles have two ways for transportation: the apoplastic and symplastic. For cell-to-cell transportation occurs through plasmodesmata. Apoplastic transport includes movement through the extracellular spaces (non-living) whereas symplastic transport includes cell membrane and plasmodesmata (Sattelmacher, 2001). For radial movement of nanoparticles, apoplastic movement is very important and they reach the xylem cylinder. Accumulation of some nano materials at the Casparian strip was observed, support the movement through xylem cylinder (Lv et al., 2015; Sun et al., 2014). After reaching to xylem, they move upwards through transpiration streams (Sun et al., 2014; Wang et al., 2012). In the cytoplasm, the NPs interact and may bind to different organelles in various ways to further interact with the plant metabolism in both positive as well as negative ways (Jia et al., 2005). Although the exact mechanism of uptake and translocation of nanoparticles is still under investigation.

7.3 EFFECT OF NANOPESTICIDES ON PLANTS

7.3.1 EFFECT ON GROWTH PARAMETERS

The primary source of nanoparticle exposure in plants is through the application of nanofertilizers and nanopesticides. Nanoparticles can interact with plants in various ways, influencing growth and development. While "growth" is a common term applied to changes in organisms, in plants, its meaning varies depending on the scale and context. At the meristematic tissue level, growth refers to the formation of new organs through cell division; at the organ level, it signifies an increase in cell size; and at a spatial or temporal level, it can be defined as net primary production (Hilty et al., 2021). The effects of nanoparticles on plant growth are further discussed in the following subsection.

7.3.1.1 EFFECT ON BIOMASS

The weight of living plant material that is present above and below a specific unit of ground surface area at any given time is known as plant biomass. The majority of agricultural and plant biomass is lignocellulosic and is fundamentally made up of highly crystalline cellulose, hemicellulose, and lignin (Singh and Sarma, 2022). The use of nanoparticles is increased day by day and impact on plant growth and development is reflected into biomass. Impacts on biomass of some economically important crop were listed in Table 7.1. During the plant growth, it absorbs excessive quantity of essential and non-essential elements. Ions and water is important carrier to cross the channels and aquaporin of plant cell due to its smaller size and diameter and enters in the plasma membrane. The plants treated with nanoparticles also works with same principle as relatively small size and large surface area. The rice, wheat and maize are being the most important staple crops all over the world so becomes the model plants for the research and evaluating the application of nanopesticides. It is clear in reports that silver nanoparticles are at different concentration imparts negative effect on seedling growth.

The influence of ZnO NPs at 25 mg/L has increased the seedling length in rice plant and also show positive results on dry weight biomass and chlorophyll content (Itroutwar et al., 2020). Nanoparticles from iron metal are proven to be beneficial for the all the three plants by increasing chlorophyll content, seedling length and shoot root biomass. TiO_2 NP show impactful results in rice and wheat but there is variation in maize, as at low concentration (5–20 mg/L) it has reduced the chlorophyll content and dry biomass (Daghan, 2018). The results of carbon nanotubes suggest a possible link for plant hormones under CNTs treatment to promote seedlings growth in all three crop plant. The concentration range varies with the variation in plant species. The foliar applications were more effective than the soil treatments. The authors reported that the highest stimulated factors are plant height, root length and biomass at a low doses but there is toxic effect of ZnO, Fe_2O_3, and carbon nanotubes on exposure at higher doses.

7.3.1.2 EFFECT ON PLANT HORMONES

Small endogenous signaling molecules called plant growth regulators include auxins, cytokinin, gibberellins, salicylic acid (SA), ABA, ethylene, brassinosteroids, and jasmonic acid (JA). Auxin, gibberellin, and cytokinin, three of these hormones, are recognized to play significant roles in plant developmental responses; meanwhile, ABA, SA, JA, and ethylene are engaged in modulating plant defensive responses to biotic and abiotic stresses (Khan et al., 2020; Tripathi et al., 2022).

Crop productivity and quality can be enhanced by using nanoparticles as nanofertilizers nanopesticides and nanohormone, in this context, Sygenta a agrochemical company formulated The Primo MAXX (a nanoproduct that function as plant growth regulators) which provide turf grass to withstand biotic and abiotic stresses (Rani et al., 2020). Recent Studies have reported that nanoparticles could control phytohormones responses to control stresses by modulating biosynthesis or regulating signaling process and their control responses depend upon various factors such as type of nanoparticles, plant species and way of exposure, etc. (Table 7.2) (Rasool, 2022; Yue et al., 2017).

TABLE 7.1 Effects of Various Nanoparticles on Biomass of Economically Important Crops

Experimented Plant	Type of Nanoparticles	Parameters	Concentration	Result	References
Zea mays	Ag	Shoot and root biomass	2.5 mg/kg	Decreased	Cao et al., (2017)
	ZnO	Dry weight of shoot and root Chlorophyll content	0.5 ppm	Increased	Adhikari et al., (2015)
	Fe_2O_3	Root length Chlorophyll content	50 mg/kg Foliar spray	Increased	Fathi et al., (2017)
	Carbon nanotube	Dry weight shoot and root Chlorophyll content	100 mg/L	Increased	Hu et al., (2021)
	TiO_2	Chlorophyll content Dry weight shoot and root Shoot nutrient uptake (NPK)	5–20 mg/L	Reduced Reduced Increased	Daghan, (2018)
Oryza sativa	Ag	Dry weight root and shoot Seedling length	200 mg/L	Decreased	Abbas et al., (2019)
	ZnO	Dry weight root and shoot Seedling length	25 mg/L	Increased	Itroutwar et al., (2020)
	Fe_2O_3	Seedling length Chlorophyll content	50 mg/L	Increased	Li et al., (2021)
	Carbon nanotube	Dry weight root and shoot Seedling length Chlorophyll content	20 mg/L	Increased	Zhang et al., (2017)
	TiO	Dry weight root and shoot Seedling length Chlorophyll content	500 mg/l	Increased	Arshad et al., (2021)
Triticum aestivum	Ag	Dry weight root and shoot Seedling length	10 mg/L	Decreased	Vannini et al., (2014)
	ZnO	Dry weight root and shoot Seedling length	50ppm	Increased	Rawat et al. (2018)
	Fe_3O_4	Dry weight root and shoot Seedling length	500 mg/L	Increased	Feng et al., (2022)
	Carbon nanotube	Dry weight root and shoot Seedling length	80 mg/L	Increased	Chang et al., (2020)
	TiO_2	Dry weight root and shoot Seedling length	50 ppm	Increased	Rawat et al. (2018)

TABLE 7.2 Effects of Various Nanoparticles on Plant Growth Hormones

Type of Nanoparticles	Exposure	Effects on Plant Hormone	References
Se	Foliar spray	Enhancement of IAA and Abscisic acid and improved growth and yield of Strawberry plant against saline stress.	Zahedi et al., (2019)
Ag	Seed soaking	Elongation of Root and regulate the transcriptome of JAL gene that regulate auxin biosynthesis in *Eruca sativa*	Vannini et al., (2013)
Ag	Seed Foliage	Induced the expression of two gene NCED and RD22 that are responsible for ABA synthesis Induced ABA content in Wheat plant under Salinity stress	Syu et al., (2014) Wahid et al., (2020)
Cu	Root exposure	Change in root morphology of wheat plant by modifying auxin transporter genes	Zhang et al., (2018)
TiO_2		Promote root growth, increase in the of auxin expression, biosynthesis and transporter genes.	Wei et al., (2020)
Ag & Cu	Foliar application	Increased GA_3 content in leaves of *Annona muricata*	Jonapá-Hernández et al., (2020)
TiO_2 CeO_2	Seed	Upregulated Cytokinins regulating gene	Tumburu et al., (2015)
ZnO SiO_2 ZnO/SiO_2	Foliar	Under ZnO nanoparticles treatment, ethylene signaling, and biosynthesis gene were upregulated whereas cytokinin's signaling gene (ARRJ and ARR15) were upregulated under SiO_2 nanoparticles in *Arabidopsis* plant	Azhar et al., (2021)

7.3.2 EFFECT OF NANOPESTICIDES ON PHOTOSYNTHESIS

Photosynthesis is a key energy transformation and metabolic reactions in plants that convert light energy into chemical energy. This process is complete into two steps: the first step is known as light dependent reaction and second one is known as dark reaction (Light independent). Every life on earth gets energy directly or indirectly through photosynthesis process (Schlesinger, 2020). Only 2 to 4% of solar energy is used by plants for transformation into chemical energy. In recent years, researchers have been using nano-technological tools such as nano formulation or nanopesticides to investigate their role in plant growth, metabolism and especially photosynthesis that affect plant/crop productivity. It has been investigated that nanoparticles can enhance or slow down the photosynthesis process, but the effect depends on various factors such as metals used for nano synthesis, plants at species level, uptake and method of applications. Structure of chloroplast, CO_2 concentration, thylakoid protein and photosynthetic enzymes (RuBisCO, PEP and carbonic anhydrase, etc.) are the main influencing factors that affect photosynthesis.

Nano materials either can show adverse effects on photosynthesis by blocking electron transport chain, suppressing the activity of photosynthetic enzymes or can boost up by improving the capacity of light harvesting system (Jampílek & Kráľová, 2019; Shweta et al., 2016). Metal and metal oxide such as Ag, Au, Zn, Pt, Cu, TiO_2, CuO, CeO_2, and ZnO are mostly used and distributed nanoparticles all over the world and have been explored to know the impacts on photosynthetic machinery (Table 7.3). Foliar application of metallic nanoparticles showed significant increase in chlorophyll content of treated plants by synthesizing more LHC to capture light (Figure 7.2) (Kataria et al., 2019).

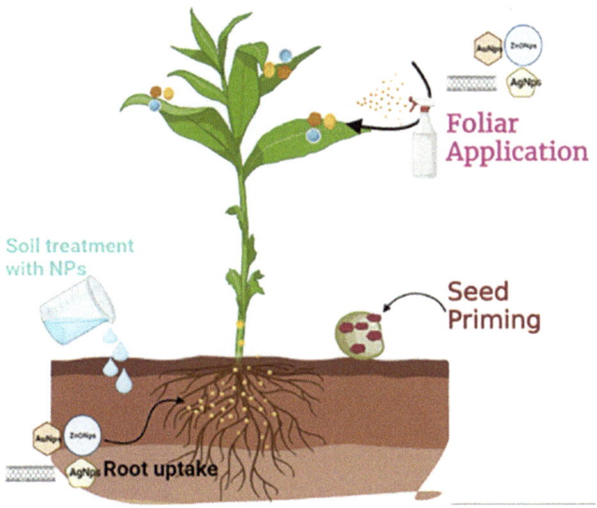

FIGURE 7.2 Various methods of application of nanoparticles.

Silver nanoparticles (AgNPs) are among the most widely used metallic nanoparticles in biosensors and pesticides due to their unique physical and chemical properties, including high surface area, thermal and electrical conductivity, and catalytic activity. However, various studies have observed negative impacts on photosynthesis, such as decreased chlorophyll content, impaired PSII efficiency, and inhibited electron transport in plants (Huang et al., 2016; Matorin et al., 2013; Queiroz et al., 2016). Zinc (Zn) and titanium dioxide (TiO_2) nanoparticles have shown both negative and positive effects on the photosynthesis process. Studies by Liu et al. (2015) and Nayan et al. (2016) reported decreases in carotenoid and chlorophyll content in *Zea mays* and *Brassica juncea* due to Zn and ZnO nanoparticles, respectively. In contrast, a threefold increase in total chlorophyll content in *Cyamopsis tetragonoloba* was observed following foliar application of ZnO nanoparticles (Raliya & Tarafdar, 2013).

Studies have shown that titanium dioxide (TiO_2) nanoparticles negatively impact the photosynthetic process, both structurally and functionally, in various microalgal species such as *Scenedesmus* and *Chlorella* sp. (Barreto & Lombardi, 2016; Li et al., 2020; Middepogu et al., 2018). In contrast, in agriculturally important crops like *Cicer arietinum* and *Vicia faba*, TiO_2 nanoparticles have been observed to enhance photosynthesis by increasing Rubisco activity or promoting electron transport (Sarmast & Salehi, 2016). The effects of gold (Au), cerium dioxide (CeO_2), and cobalt (Co) nanoparticles on photosynthesis remain unclear.

Non-metallic nanoparticles, such as single-walled carbon nanotubes (SWCNT) and silicon dioxide (SiO_2), are noted for their lower oxidative potential and biocompatibility. For example, SiO_2 nanoparticles have been applied as foliar fertilizers to detoxify maize grown in aluminum-contaminated soils (de Sousa et al., 2019). Beyond detoxification, these nanoparticles have been shown to significantly increase photosynthesis rates, stomatal conductance, and pigment synthesis in *Crataegus* species (Ashkavand et al., 2015). Additionally, SWCNTs embedded in chloroplasts have demonstrated positive effects on the photosynthesis process by enhancing electron transport and the production of signaling molecules such as nitric oxide (NO), which further promotes photosynthesis (Giraldo et al., 2014). Similarly, Wang et al. (2018) reported a beneficial effect of carbon dots, a type of carbon nanomaterial, on photosynthesis in mung beans.

TABLE 7.3 Effect of Nanoparticles on Photosynthesis

Nano Materials	Size (In nm)	Way of Exposure	Tested Plants	Effects	References
ZnO	24 ± 3	Soil amended (0-400 mg/kg)	*Coriandrum sativum*	Increase photosynthesis and decrease lipid peroxidation	Reddy Pullagurala et al., (2018)
	≤30	Foliar application (200-800 mg/l)	*Solanum lycopersicum*	Decreased in content of chlorophylls a and b, reduction in photosynthetic efficiency	Wang et al., (2018)
	≤30	Foliar Application	*Crocus sativus*	Increase chlorophyll content and POX & CAT activity	Rostami et al., (2019)
		Seed Priming	*Triticum aestivum*	Increase in content of photosynthetic pigments and overall efficiency	Rai-Kalal and Jajoo, (2021)
		Foliar application (50 mg/l)	*Oryza sativa*	Positive effect on photosynthesis parameters such as chlorophyll content, high fluorescence	Faizan et al., (2021)
Ag	50-100	Incubation	*Chlorella vulgaris*	Negative effect on photosynthesis	Hazeem et al., (2019)
	50	Incubation	*Lemna gibba*	Negative effect	Dewez et al., (2018)
				Decrease chlorophyll content and block PSII	
TiO2	5-10	Incubation	*Oryza sativa*	Increased photosynthetic rate	Zhang et al., (2020)
	Average size 14nm	Foliar spray	*Vetiveria zizanioides*	Increased total chlorophyll content, photochemical efficiency of PSII	Shabbir et al., (2019)
		Foliar spray	*Vitis vinifera*	Negative effect photosynthesis	Teszlák et al., (2018)
	12±3.5	Incubated with concentration of o.1 to 20 mg/l	*Chlorella pyrenoidosa*	Negatively affect photosynthesis by destroying lamellar structure of thylakoids and damaging reaction Centre PSII	Middepogu et al., (2018)
FeO	54±1	Soil application	*Arabidopsis thaliana*	Increased photosynthetic rate (CO_2 assimilation rate, stomatal conductance)	Yoon et al., (2019)
Fe_2O_3	Coated with critic acid (20 nm)	Foliar spray	*Glycine max*	Positive effect	Alidoust and Isoda, (2013)

7.3.3 EFFECTS ON SEED GERMINATION

Seed is the initial factor for the future plant. Thus, peculiarity of seed is important in crop development, whereas the impact of other inputs, like fertilizer doses and irrigation methods which are also determine the seed quality. Quality of a seed is defined as high germination rate, the genetic purity, and adequate moisture content, physical, physiological and genetic health and free from any infestation of diseases insects or pests. Likewise, physiological quality of a seed is evaluated by germination percentage, seed vigor, which can be targeted using various seed treatment methods such washing, priming, scarification, stratification, seed pelleting, etc.

Nanoparticles have been used as potential plant growth matter and have ability to provide effective fertilizers, pesticides, and nutrients for seed growth as well as seed dormancy breaker. Nanoparticles may enter into seed via different pathways like by penetrating seed coat (Radchuk & Borisjuk, 2014) or by creating new pores through up regulation of aquaporin genes (Miralles et al., 2012; Obroucheva, 2013). In recent study showed that ZnO and TiO_2 nanoparticles enter physically through seed coat and stimulating embryo differentiation by upregulating enzymes that help in breaking of seed dormancy (Miralles et al., 2012). There are several experiments showing benefits of nanoparticles in environmental and agricultural applications, especially Improvement of seed quality parameters by increasing water absorption, by supporting seed enzyme activity and imparting defense mechanisms against natural threats and pathogens (Table 7.4).

The nanoparticles decompose as ions in the soil that would have been picked up by plants and seeds as nourishments to enhance their beneficial effects. In addition, higher concentration of these nanoparticles application in soil, especially synthetic ones nanoparticles can also adversely affect plants parameters like retardation of germination index, reduced root length and shoot length, seedling index, induces stress, and inhibits beneficial-microorganisms growth.

7.3.4 EFFECT ON ANTIOXIDANT ACTIVITY

Reactive oxygen species (ROS), such as superoxide anion radicals (O_2) hydrogen oxides (H_2O_2), hydroxyl radicals (OH), and O_2 (singlet oxygen) are produced by a variety of biotic and abiotic stresses as well as cellular

TABLE 7.4 Effect of Various Nanoparticles on Seed Germination

SL. No.	Host Species	Nanoparticle	Subjection Mode	Exposure Conc.	Result /Effect	References
1	*Citrullus lanatus*	Ag (AgNP)	Suspension	31.3 ppm	Seed priming enhanced seed germination, growth, and yield	Acharya et al., (2020)
2	*Allium cepa*	Ag–au	Suspension	31.3 ppm	Germination stimulators such as aminobutyric acid and zeatin are enhanced in onion seeds and seedlings.	Acharya et al., (2020)
3	Ryegrass, barley, and flax	Ag & Fe	Suspension	10 mg/l and 250 mg/l	Inhibitory effects in aqueous suspensions at 250 mg l^{-1} for Fe and 10 mg l^{-1} for Ag.	El-Temsah and Joner, (2012)
4	*Oryza sativa*	Ag	Suspension	20 ppm	Increased shoot, root length with effective rate of seed germination.	Gupta et al., (2018)
5	*Oryza sativa*	Ag	Suspension	5-10 ppm	Improvement in seed germination and seedling vigor.	Mahakham et al., (2017)
6	*Raphanus sativus*	Ag	Suspension	500 µg/ml	A phyto-stimulant effect on seed germination.	Segura et al., (2020)
7	*Triticum aestivum*	Ag	Suspension	100-150 mg/l	Significant reduction in germination with 100–150 mg l^{-1} and no effect at lower concentration.	Y. Singh et al., (2020)
8	*Lycopersicon esculentum*	Ag &TiO_2	Suspension	100 mg/kg	Phyto-toxic effect on plant growth.	Song et al., (2013)
9	*Cajanus cajan*	Ag	Suspension	1.2 nm	Promoted germination percent under fluoride stress condition.	Yadu et al., (2018)
10	*Raphanus sativus*	Ag	Suspension	125 mg/l	96% germination growth	Zuverza-Mena et al., (2016)
11	*Phaseolus vulgaris*	Al	Suspension	1.6 mg/ml	No major effects on seeds germination, on radicle and shoot elongation.	Belhamel et al., (2020)
12	*Raphanus sativus*	Ce	Mixed with soil	62.5–125 mg/kg 500 mg/kg.	Slight increment in seed germination %; but 500 mg/kg retard the seed germination.	Corral-Diaz et al., (2014)
13	*Vigna radiata* and *Cajanus cajan*	MgO	Seed primimg	5 mg/l	100% seed germination rate.	Rani et al., (2020)

TABLE 7.4 *(Continued)*

SL. No.	Host Species	Nanoparticle	Subjection Mode	Exposure Conc.	Result /Effect	References
14	*Phaseolus vulgaris*	Iron oxide	Suspension	1-10 mg/l	95–97% Germination rate and elongation of radical in seedling.	Duran et al., (2018)
15	*Citrullus lanatus*	Iron oxide	Suspension	160 mg/l	Fe-NPs was safer compared to its bulk counterparts ($fecl_3$ and fe_2o_3) and had no toxic impact on seed germination	Kasote et al., (2019)
16	*Zea mays*	Iron oxide nanoparticles (γ-fe_2o_3 nps)	Suspension	50 mg/l	20 mg/l of NPs have significant results in root elongation by 11.5%, and germination index but 100 mg/l dose decreased the germination index.	Li et al., (2016)
17	*Raphanus sativus*	Iron oxide	Suspension	1.0 mg/ml	Fe_2o_3-nps retarded the root length and seed germination in radish	Saquib et al., (2016)
18	*Vigna radiata*	Iron oxide	Suspension	600–750 mg/l	Decreased germination rate at 600–750 mg/l	Sun et al., (2019)
19	*Ocimum basilicum*	Hydrophobic and hydrophilic nano-TiO_2	Mixture with Soil	750 mg/kg	Both types Hydrophobic and hydrophilic nano-tio_2 significantly reduced seed germination by 41% and 59%, respectively.	Tan et al., (2017)
20	*Solanum lycopersicum*	ZnO	Suspension	1.2 mm	Lower concentration of Nps exhibited maximum germination	Singh et al., (2016)
21	*Triticum aestivum*	ZnO	Suspension	250 mg/l	High seedling emergence and seed germination.	Singh et al., (2019)
22	Green and red beans	Zno fabricated with flower extract of *Canna indica*	Suspension	10 mg/l	Germination ratio is 100%.	Nguyen et al., (2021)
23	*Zea mays*	ZnO	Suspension	0.1 mg/ l	At 30 °c, 0.1 mg l– increased the growth by 50%. 20c reduced the seed germination.	López-Moreno et al., (2017)
24	*Lactuca sativa*	Carbon	Suspension	0.3%	Improved seed germination under salinity condition	Baz et al., (2020)
25	*Oryza sativa*	Carbon	Suspension	100ug/ml 150ug/ml	Promotion of seed germination and root growth . Decreased the growth.	Jiang et al., (2014)

oxidation processes in Chloroplast, Mitochondria and Peroxisome. There are also secondary sites for production of ROS such as Endoplasmic reticulum, cell membrane and apoplast (Das & Roychoudhury, 2014). ROS attack all macro molecules like oxidation of protein, lipids, carbohydrates and DNA resulting into oxidative stress and ultimately cause the death of cell (Karuppanapandian et al., 2011).

Various antioxidative defense systems that include enzymatic and non-enzymatic molecules scavenge that ROS. The enzymatic machinery comprises of superoxide dismutase (SOD), peroxidase (ascorbate and guaiacol), and catalase (CAT) and non-enzymatic machinery is comprises of ascorbic acid, carotenoids, phenolics and flavonoids (Gill and Tuteja, 2010). Nanoparticles may impacts on antioxidant enzymes in both positive and negative way. Various studies showed that metallic nanoparticles enhance tolerance against abiotic stress by activation production of antioxidant enzymes and their uptake in plants (Table 7.5) (Tripathi et al., 2013; Tripathi, Shweta, et al., 2017). SOD, APX (Ascorbate peroxidase) and GST (Glutathione-S-transferase) antioxidant enzymes were up regulated in silver nanoparticles treated *Oryza sativa* roots (Mirzajani et al., 2014) while in *Pisum sativum* plant; these nanoparticles inhibit glutathione reductase (GR) and dehydroascorbate reductase (DHAR) were down regulated (Tripathi, Singh, et al., 2017). Catalase (CAT), an enzyme that protects cells from oxidative damage, was upregulated in copper oxide nanoparticles-treated wheat plants (Dimkpa et al., 2012).

However, it is difficult to generalize the impacts of nanoparticles on antioxidant activity of plants but it is necessary to vast our knowledge and understanding on the fate of nanoparticles and their relevant conditions and concentration in the context of ROS, antioxidant activity, signaling and tolerance of oxidative stress in plants.

7.4 CONCLUSION AND FUTURE PERSPECTIVE

Nano protectant of plants and nanopesticides are emerging as latest nano technology for replacing the current use of pesticides with many advantage such as effectiveness, sustainability and targeted delivery and reduction in the amount of active ingredient. Many formulations techniques such as nano-capsulation, nano-emulsion, inorganic nanoparticles (containing metals and metal oxides), nano-clay and nano-carbon tubes are practicing

TABLE 7.5 Effects of Nanoparticles on Antioxidant Activity of Plants

Nano Particles	Mode of Application	Plant	Effect	References
$Cu(OH)_2$	Foliar Spray 0.18-18 mg/plant	*Spinacia oleracea*	Induced alternation in metabolites and reduced antioxidant metabolites	Zhao, Huang, et al., (2017)
$Cu(OH)_2$	Foliar spray 10-100 mg/l	*Zea mays*	Gene expression of antioxidant enzymes (POD, GST) is increased below 100 mg con., phenolic compounds increase	Zhao, Hu, et al., (2017)
TiO_2	Growth medium (10-2000 mg/l)	*Lemna minor*	POD, SOD and CAT enzymes activity is increased below of 200 mg conc.	Song et al., (2012)
TiO_2	Suspension medium (0.05 -10 mg/l)	*Spirodela polyrrhiza*	The activity of SOD is increased by increasing conc. of nanoparticles but no significant result is showed by other enzymes	Movafeghi et al., (2018)
ZnO	Suspension (200-800 mg/m^{3})	*Solanum lycopersicum*	Decreased growth of root, shoot and chlorophyll a & b content. SOD, CAT and APX activity enhanced in conc. dependent manner.	Wang et al., (2018)
Ag	Seedling treatment 10-40 ppm	*Oryza sativa*	Promote the growth of seedling and upregulated the expression of CAT, APX and GR antioxidant enzymes gene	Gupta et al., (2018)
Ag	Seedling treatment (25-400ppm)	*Brassica juncea*	Promoted the overall growth and chlorophyll content, decreased the level of hydrogen peroxide and malondialdehyde by enhancing the activity of APX and GPX	Sharma et al., (2012)
Ag & Au	Medium includes (2-50 mg/m^3)	*Lavandula angustifolia*	Increased activity of APX and SOD were observed and the reaction was conc. dependent.	Jadczak et al., (2020)
Carbon NanoTubes (CNTs) and Carbon nanoparticles (CNPs)	Seedling treatment	*Zea mays*	Enhanced the both germination and growth of seeds under Cu stress, also activated SOD, POD and CAT enzymes. CNPs is more effective than to CNTs	Xin et al., (2022)

to develop innovative application of nanotechnology in agriculture sector so that these products can be utilized to enhance the effectuality and sustainability of traditional pesticides and enhance their environmental safety profiles. However, owing to their distinctive size and properties an abundant researches have been done to know their effects/impacts on yield and production of economically important plant. There is little knowledge on their interaction with plant and environment in context to climate or available condition. The investigations on nanoparticles, which are crucial for plants, is still in its infancy, as this chapter makes clear; more thorough studies are required to comprehend the biochemical, physiological, and molecular mechanisms of plants in connection to NPs. Additionally, in-depth research is required to examine the way NPs work, how they interact with biomolecules, and how they affect the control of gene expression in plants.

KEYWORDS

- **nanoagrochemicals**
- **nanofertilizers**
- **nanohormones**
- **nanoparticles**
- **nanopesticides**
- **nano-scale**
- **nanotechnology**
- **plasmodesmata**

REFERENCES

Abbas, Q., Liu, G., Yousaf, B., Ali, M. U., Ullah, H., & Ahmed, R. (2019). Effects of biochar on uptake, acquisition and translocation of silver nanoparticles in rice (*Oryza sativa* L.) in relation to growth, photosynthetic traits and nutrients displacement. *Environmental Pollution, 250,* 728–736. https://doi.org/10.1016/j.envpol.2019.04.083.

Acharya, P., Jayaprakasha, G. K., Crosby, K. M., Jifon, J. L., & Patil, B. S. (2020). Nanoparticle-mediated seed priming improves germination, growth, yield, and quality of watermelons (*Citrullus lanatus*) at multi-locations in Texas. *Scientific Reports, 10*(1), 5037. https://doi.org/10.1038/s41598-020-61696-7.

Acharya, P., Jayaprakasha, G. K., Semper, J., & Patil, B. S. (2020). 1H nuclear magnetic resonance and liquid chromatography coupled with mass spectrometry-based metabolomics reveal enhancement of growth-promoting metabolites in onion seedlings treated with green-synthesized nanomaterials. *Journal of Agricultural and Food Chemistry, 68*(46), 13206–13220. https://doi.org/10.1021/acs.jafc.0c00817.

Adhikari, T., Kundu, S., Biswas, A. K., Tarafdar, J. C., & Subba Rao, A. (2015). Characterization of zinc oxide nanoparticles and their effect on growth of maize (*Zea mays* L.). *Journal of Plant Nutrition, 38*(10), 1505–1515. https://doi.org/10.1080/01904167.2014.992536.

Alidoust, D., & Isoda, A. (2013). Effect of γFe2O3 nanoparticles on photosynthetic characteristics of soybean (*Glycine max* (L.) Merr.): Foliar spray versus soil amendment. *Acta Physiologiae Plantarum, 35*(12), 3365–3375. https://doi.org/10.1007/S11738-013-1369-8.

Arshad, M., Nisar, S., Gul, I., Nawaz, U., Irum, S., Ahmad, S., Sadat, H., Mian, I. A., Ali, S., Rizwan, M., Alsahli, A. A., & Alyemeni, M. N. (2021). Multi-element uptake and growth responses of rice (*Oryza sativa* L.) to TiO2 nanoparticles applied in different textured soils. *Ecotoxicology and Environmental Safety, 215,* 112–149. https://doi.org/10.1016/j.ecoenv.2021.112149.

Ashkavand, P., Tabari, M., Zarafshar, M., Tomášková, I., & Struve, D. (2015). Effect of SiO2 nanoparticles on drought resistance in hawthorn seedlings. *Forest Research Papers, 76*(4), 350–359. https://doi.org/10.1515/frp-2015-0034.

Asif, M., Islam, S., Malik, M. A., Dar, Z. M., Masood, A., Shafi, S., Rashid, B., Sidique, S., & Asif, M. (2021). Nanopesticides application in agriculture and their impact on environment. In *Agricultural Development in Asia—Potential Use of Nano-Materials and Nano-Technology* (pp. 1–15). Intech Open. https://doi.org/10.5772/intechopen.100690.

Azhar, B. J., Noor, A. S. M. A., Zulfiqar, A. L. V. E. E. N. A., Zeenat, A., Shakeel, A., Chishti, I., & Shakeel, A. N. (2021). Effect of ZnO, SiO2, and composite nanoparticles on *Arabidopsis thaliana* and involvement of ethylene and cytokinins signaling pathway. *Pak J Bot, 53*(2), 437–446.

Barreto, D. M., & Lombardi, A. T. (2016). Environmentally relevant concentrations of TiO_2 nanoparticles affected cell viability and photosynthetic yield in the chlorophyceae *Scenedesmus bijugus*. *Water, Air, & Soil Pollution, 227*(12), 450. https://doi.org/10.1007/s11270-016-3139-x.

Baz, H., Creech, M., Chen, J., Gong, H., Bradford, K., & Huo, H. (2020). Water-soluble carbon nanoparticles improve seed germination and post-germination growth of lettuce under salinity stress. *Agronomy, 10*(8). https://doi.org/10.3390/agronomy10081192.

Belhamel, C., Boulekbache–Makhlouf, L., Bedini, S., Tani, C., Lombardi, T., Giannotti, P., Madani, K., Belhamel, K., & Conti, B. (2020). Nanostructured alumina as seed protectant against three stored-product insect pests. *Journal of Stored Products Research, 87,* 101607. https://doi.org/10.1016/j.jspr.2020.101607.

Bollag, J. M., Myers, C. J., & Minard, R. D. (1992). Biological and chemical interactions of pesticides with soil organic matter. *Science of the Total Environment, 123,* 205–217. https://doi.org/10.1016/0048-9697(92)90146-J.

Cao, J., Feng, Y., He, S., & Lin, X. (2017). Silver nanoparticles deteriorate the mutual interaction between maize (*Zea mays* L.) and arbuscular mycorrhizal fungi: A soil microcosm study. *Applied Soil Ecology, 119,* 307–316. https://doi.org/10.1016/j.apsoil.2017.04.011.

Chang, X., Song, Z., Xu, Y., & Gao, M. (2020). Effects of carbon nanotubes on growth of wheat seedlings and Cd uptake. *Chemosphere, 240,* 124931. https://doi.org/10.1016/j.chemosphere.2019.124931.

Chaud, M., Souto, E. B., Zielinska, A., Severino, P., Batain, F., Oliveira-Junior, J., & Alves, T. (2021). Nanopesticides in agriculture: Benefits and challenges in agricultural productivity, toxicological risks to human health and the environment. *Toxics, 9*(6). https://doi.org/10.3390/toxics9060131.

Conley, D. J., Paerl, H. W., Howarth, R. W., Boesch, D. F., Seitzinger, S. P., Havens, K. E., Lancelot, C., & Likens, G. E. (2009). Controlling eutrophication: Nitrogen and phosphorus. *Science, 323*(5917), 1014–1015. https://doi.org/10.1126/science.1167755.

Corral-Diaz, B., Peralta-Videa, J. R., Alvarez-Parrilla, E., Rodrigo-García, J., Morales, M. I., Osuna-Avila, P., Niu, G., Hernandez-Viezcas, J. A., & Gardea-Torresdey, J. L. (2014). Cerium oxide nanoparticles alter the antioxidant capacity but do not impact tuber ionome in *Raphanus sativus* (L). *Plant Physiology and Biochemistry, 84,* 277–285. https://doi.org/10.1016/j.plaphy.2014.09.018.

Daghan, H. (2018). Effects of TiO2 nanoparticles on maize (*Zea mays* L.) growth, chlorophyll content and nutrient uptake. *Applied Ecology and Environmental Research, 16,* 6873–6883. https://doi.org/10.15666/aeer/1605_68736883.

Das, K., & Roychoudhury, A. (2014). Reactive oxygen species (ROS) and response of antioxidants as ROS-scavengers during environmental stress in plants. *Frontiers in Environmental Science, 2.* https://www.frontiersin.org/articles/10.3389/fenvs.2014.00053.

De Sousa, A., Saleh, A. M., Habeeb, T. H., Hassan, Y. M., Zrieq, R., Wadaan, M. A. M., Hozzein, W. N., Selim, S., Matos, M., & AbdElgawad, H. (2019). Silicon dioxide nanoparticles ameliorate the phytotoxic hazards of aluminum in maize grown on acidic soil. *The Science of the Total Environment, 693,* 133636. https://doi.org/10.1016/j.scitotenv.2019.133636.

Delfani, M., Baradarn Firouzabadi, M., Farrokhi, N., & Makarian, H. (2014). Some physiological responses of black-eyed pea to iron and magnesium nanofertilizers. *Communications in Soil Science and Plant Analysis, 45*(4), 530–540. https://doi.org/10.1080/00103624.2013.863911.

Dimkpa, C. O., McLean, J. E., Latta, D. E., Manangón, E., Britt, D. W., Johnson, W. P., Boyanov, M. I., & Anderson, A. J. (2012). CuO and ZnO nanoparticles: Phytotoxicity, metal speciation, and induction of oxidative stress in sand-grown wheat. *Journal of Nanoparticle Research, 14*(9), 1125. https://doi.org/10.1007/s11051-012-1125-9.

Duran, N. M., Medina-Llamas, M., Cassanji, J. G. B., de Lima, R. G., de Almeida, E., Macedo, W. R., Mattia, D., & Pereira de Carvalho, H. W. (2018). Bean seedling growth enhancement using magnetite nanoparticles. *Journal of Agricultural and Food Chemistry, 66*(23), 5746–5755. https://doi.org/10.1021/acs.jafc.8b00557.

El-Temsah, Y. S., & Joner, E. J. (2012). Impact of Fe and Ag nanoparticles on seed germination and differences in bioavailability during exposure in aqueous suspension and soil. *Environmental Toxicology, 27*(1), 42–49. https://doi.org/10.1002/tox.20610.

Etxeberria, E., Gonzalez, P., Baroja-Fernández, E., & Romero, J. P. (2006). Fluid phase endocytic uptake of artificial nano-spheres and fluorescent quantum dots by sycamore cultured cells. *Plant Signaling & Behavior, 1*(4), 196–200. https://doi.org/10.4161/psb.1.4.3142.

Faizan, M., Bhat, J. A., Hessini, K., Yu, F., & Ahmad, P. (2021). Zinc oxide nanoparticles alleviate the adverse effects of cadmium stress on *Oryza sativa* via modulation of the photosynthesis and antioxidant defense system. *Ecotoxicology and Environmental Safety, 220*. https://doi.org/10.1016/J.ECOENV.2021.112401.

Fathi, A., Zahedi, M., & Torabian, S. (2017). Effect of interaction between salinity and nanoparticles (Fe_2O_3 and ZnO) on physiological parameters of *Zea mays* L. *Journal of Plant Nutrition, 40*(19), 2745–2755. https://doi.org/10.1080/01904167.2017.1381731.

Feng, Y., Kreslavski, V. D., Shmarev, A. N., Ivanov, A. A., Zharmukhamedov, S. K., Kosobryukhov, A., Yu, M., Allakhverdiev, S. I., & Shabala, S. (2022). Effects of iron oxide nanoparticles (Fe_3O_4) on growth, photosynthesis, antioxidant activity, and distribution of mineral elements in wheat (*Triticum aestivum*) plants. *Plants, 11*(14). https://doi.org/10.3390/plants11141894.

Ghafariyan, M. H., Malakouti, M. J., Dadpour, M. R., Stroeve, P., & Mahmoudi, M. (2013). Effects of magnetite nanoparticles on soybean chlorophyll. *Environmental Science & Technology, 47*(18), 10645–10652. https://doi.org/10.1021/es402249b.

Gill, S. S., & Tuteja, N. (2010). Reactive oxygen species and antioxidant machinery in abiotic stress tolerance in crop plants. *Plant Physiology and Biochemistry, 48*(12), 909–930. https://doi.org/10.1016/j.plaphy.2010.08.016.

Giraldo, J. P., Landry, M. P., Faltermeier, S. M., McNicholas, T. P., Iverson, N. M., Boghossian, A. A., Reuel, N. F., Hilmer, A. J., Sen, F., Brew, J. A., & Strano, M. S. (2014). Plant nanobionics approach to augment photosynthesis and biochemical sensing. *Nature Materials, 13*(4), 4. https://doi.org/10.1038/nmat3890.

Gupta, S. D., Agarwal, A., & Pradhan, S. (2018). Phytostimulatory effect of silver nanoparticles (AgNPs) on rice seedling growth: An insight from antioxidative enzyme activities and gene expression patterns. *Ecotoxicology and Environmental Safety, 161*, 624–633. https://doi.org/10.1016/j.ecoenv.2018.06.023.

Hayles, J., Johnson, L., Worthley, C., & Losic, D. (2017). Nanopesticides: A review of current research and perspectives. In A. M. Grumezescu (Ed.), *New Pesticides and Soil Sensors* (pp. 193–225). Academic Press. https://doi.org/10.1016/B978-0-12-804299-1.00006-0.

Hazeem, L. J., Kuku, G., Dewailly, E., Slomianny, C., Barras, A., Hamdi, A., Boukherroub, R., Culha, M., & Bououdina, M. (2019). Toxicity effect of silver nanoparticles on photosynthetic pigment content, growth, ROS production and ultrastructural changes of microalgae *Chlorella vulgaris*. *Nanomaterials, 9*(7), 914. https://doi.org/10.3390/NANO9070914.

Hilty, J., Muller, B., Pantin, F., & Leuzinger, S. (2021). Plant growth: The what, the how, and the why. *New Phytologist, 232*(1), 25–41. https://doi.org/10.1111/nph.17610.

Hu, Y., Zhang, P., Zhang, X., Liu, Y., Feng, S., Guo, D., Nadezhda, T., Song, Z., & Dang, X. (2021). Multi-wall carbon nanotubes promote the growth of maize (*Zea mays*) by regulating carbon and nitrogen metabolism in leaves. *Journal of Agricultural and Food Chemistry, 69*(17), 4981–4991. https://doi.org/10.1021/acs.jafc.1c00733.

Huang, J., Cheng, J., & Yi, J. (2016). Impact of silver nanoparticles on marine diatom *Skeletonema costatum*. *Journal of Applied Toxicology, 36*(10), 1343–1354. https://doi.org/10.1002/jat.3325.

Itroutwar, P. D., Govindaraju, K., Tamilselvan, S., Kannan, M., Raja, K., & Subramanian, K. S. (2020). Seaweed-based biogenic ZnO nanoparticles for improving agro-morphological characteristics of rice (*Oryza sativa* L.). *Journal of Plant Growth Regulation, 39*(2), 717–728. https://doi.org/10.1007/s00344-019-10012-3.

Jadczak, P., Kulpa, D., Drozd, R., Przewodowski, W., & Przewodowska, A. (2020). Effect of AuNPs and AgNPs on the antioxidant system and antioxidant activity of lavender (*Lavandula angustifolia* Mill.) from *in vitro* cultures. *Molecules, 25*(23), 5511. https://doi.org/10.3390/molecules25235511.

Jampílek, J., & Kráľová, K. (2019). Impact of nanoparticles on photosynthesizing organisms and their use in hybrid structures with some components of photosynthetic apparatus. In R. Prasad (Ed.), *Plant Nanobionics: Volume 1, Advances in the Understanding of Nanomaterials Research and Applications* (pp. 255–332). Springer International Publishing. https://doi.org/10.1007/978-3-030-12496-0_11.

Jia, G., Wang, H., Yan, L., Wang, X., Pei, R., Yan, T., Zhao, Y., & Guo, X. (2005). Cytotoxicity of carbon nanomaterials: Single-wall nanotube, multi-wall nanotube, and fullerene. *Environmental Science & Technology, 39*(5), 1378–1383. https://doi.org/10.1021/es048729l.

Jiang, Y., Hua, Z., Zhao, Y., Liu, Q., Wang, F., & Zhang, Q. (2014). The effect of carbon nanotubes on rice seed germination and root growth. In T. C. Zhang, P. Ouyang, S. Kaplan, & B. Skarnes (Eds.), *Advances in Plant Sciences* (Vol. 250, pp. 1207–1212). Springer Berlin Heidelberg. https://doi.org/10.1007/978-3-642-37922-2_129.

Jonapá-Hernández, F., Gutiérrez-Miceli, F., Santos-Espinosa, A., Ruíz-Lau, N., Ruíz-Valdiviezo, V., Valdez-Salas, B., & González-Mendoza, D. (2020). Foliar application of green nanoparticles in *Annona muricata* L. plants and their effects in physiological and biochemical parameters. *Biocatalysis and Agricultural Biotechnology, 28*, 101751. https://doi.org/10.1016/j.bcab.2020.101751.

Kah, M., Beulke, S., Tiede, K., & Hofmann, T. (2013). Nanopesticides: State of knowledge, environmental fate, and exposure modeling. *Critical Reviews in Environmental Science and Technology, 43*(16), 1823–1867. https://doi.org/10.1080/10643389.2012.671750.

Karuppanapandian, T., Wang, H. W., Prabakaran, N., Jeyalakshmi, K., Kwon, M., Manoharan, K., & Kim, W. (2011). 2,4-dichlorophenoxyacetic acid-induced leaf senescence in mung bean (*Vigna radiata* L. Wilczek) and senescence inhibition by co-treatment with silver nanoparticles. *Plant Physiology and Biochemistry, 49*(2), 168–177. https://doi.org/10.1016/j.plaphy.2010.11.007.

Kashyap, P. L., Xiang, X., & Heiden, P. (2015). Chitosan nanoparticle-based delivery systems for sustainable agriculture. *International Journal of Biological Macromolecules, 77*, 36–51. https://doi.org/10.1016/j.ijbiomac.2015.02.039.

Kasote, D. M., Lee, J. H. J., Jayaprakasha, G. K., & Patil, B. S. (2019). Seed priming with iron oxide nanoparticles modulates antioxidant potential and defense-linked hormones in watermelon seedlings. *ACS Sustainable Chemistry & Engineering, 7*(5), 5142–5151. https://doi.org/10.1021/acssuschemeng.8b06013.

Kataria, S., Jain, M., Rastogi, A., Živčák, M., Brestic, M., Liu, S., & Tripathi, D. K. (2019). Role of nanoparticles on photosynthesis: Avenues and applications. In D. K. Tripathi, P.

Ahmad, S. Sharma, D. K. Chauhan, & N. K. Dubey (Eds.), *Nanomaterials in Plants, Algae and Microorganisms* (pp. 103–127). Academic Press. https://doi.org/10.1016/B978-0-12-811488-9.00006-8.

Khan, N., Bano, A., Ali, S., & Babar, M. A. (2020). Crosstalk amongst phytohormones from planta and PGPR under biotic and abiotic stresses. *Plant Growth Regulation, 90*(2), 189–203. https://doi.org/10.1007/s10725-020-00571-x.

Li, J., Hu, J., Ma, C., Wang, Y., Wu, C., Huang, J., & Xing, B. (2016). Uptake, translocation, and physiological effects of magnetic iron oxide (γ-Fe_2O_3) nanoparticles in corn (*Zea mays* L.). *Chemosphere, 159*, 326–334. https://doi.org/10.1016/j.chemosphere.2016.05.083.

Li, M., Zhang, P., Adeel, M., Guo, Z., Chetwynd, A. J., Ma, C., Bai, T., Hao, Y., & Rui, Y. (2021). Physiological impacts of zero-valent iron, Fe_3O_4, and Fe_2O_3 nanoparticles in rice plants and their potential as Fe fertilizers. *Environmental Pollution, 269*, 116134. https://doi.org/10.1016/j.envpol.2020.116134.

Li, Z., Juneau, P., Lian, Y., Zhang, W., Wang, S., Wang, C., Shu, L., Yan, Q., He, Z., & Xu, K. (2020). Effects of titanium dioxide nanoparticles on photosynthetic and antioxidative processes of *Scenedesmus obliquus*. *Plants, 9*(12). https://doi.org/10.3390/plants9121748.

Liu, Q., Chen, B., Wang, Q., Shi, X., Xiao, Z., Lin, J., & Fang, X. (2009). Carbon nanotubes as molecular transporters for walled plant cells. *Nano Letters, 9*(3), 1007–1010. https://doi.org/10.1021/nl803083u.

Liu, X., Feng, Z., Zhang, F., Zhang, S., & He, X. (2006). Preparation and testing of cementing and coating nano-subnanocomposites of slow/controlled-release fertilizer. *Agricultural Sciences in China, 5*(9), 700–706. https://doi.org/10.1016/S1671-2927(06)60113-2.

Liu, X., Wang, F., Shi, Z., Tong, R., & Shi, X. (2015). Bioavailability of Zn in ZnO nanoparticle-spiked soil and the implications for maize plants. *Journal of Nanoparticle Research, 17*, 175. https://doi.org/10.1007/s11051-015-2989-2.

López-Moreno, M. L., de la Rosa, G., Cruz-Jiménez, G., Castellano, L., Peralta-Videa, J. R., & Gardea-Torresdey, J. L. (2017). Effect of ZnO nanoparticles on corn seedlings at different temperatures: X-ray absorption spectroscopy and ICP/OES studies. *Microchemical Journal, 134*, 54–61. https://doi.org/10.1016/j.microc.2017.05.007.

Lv, J., Zhang, S., Luo, L., Zhang, J., Yang, K., & Christie, P. (2015). Accumulation, speciation, and uptake pathway of ZnO nanoparticles in maize. *Environmental Science: Nano, 2*(1), 68–77. https://doi.org/10.1039/C4EN00064A.

Mahajan, P., Dhoke, S. K., & Khanna, A. S. (2011). Effect of nano-ZnO particle suspension on the growth of mung (*Vigna radiata*) and gram (*Cicer arietinum*) seedlings using plant agar method. *Journal of Nanotechnology, 2011*, e696535. https://doi.org/10.1155/2011/696535.

Mahakham, W., Sarmah, A. K., Maensiri, S., & Theerakulpisut, P. (2017). Nanopriming technology for enhancing germination and starch metabolism of aged rice seeds using Phyto synthesized silver nanoparticles. *Scientific Reports, 7*, 8263. https://doi.org/10.1038/s41598-017-08669-5.

Matorin, D. N., Todorenko, D. A., Seifullina, N. Kh., Zayadan, B. K., & Rubin, A. B. (2013). Effect of silver nanoparticles on the parameters of chlorophyll fluorescence and P700 reaction in the green alga *Chlamydomonas reinhardtii*. *Microbiology, 82*(6), 809–814. https://doi.org/10.1134/S002626171401010X.

Middepogu, A., Hou, J., Gao, X., & Lin, D. (2018). Effect and mechanism of TiO_2 nanoparticles on the photosynthesis of *Chlorella pyrenoidosa*. *Ecotoxicology and Environmental Safety, 161*, 497–506. https://doi.org/10.1016/J.ECOENV.2018.06.027.

Miralles, P., Church, T. L., & Harris, A. T. (2012). Toxicity, uptake, and translocation of engineered nanomaterials in vascular plants. *Environmental Science & Technology, 46*(17), 9224–9239. https://doi.org/10.1021/es202995d.

Mirzajani, F., Askari, H., Hamzelou, S., Schober, Y., Römpp, A., Ghassempour, A., & Spengler, B. (2014). Proteomics study of silver nanoparticles toxicity on *Oryza sativa* L. *Ecotoxicology and Environmental Safety, 108*, 335–339. https://doi.org/10.1016/j.ecoenv.2014.07.013.

Moore, M. N. (2006). Do nanoparticles present ecotoxicological risks for the health of the aquatic environment? *Environment International, 32*(8), 967–976. https://doi.org/10.1016/j.envint.2006.06.014.

Movafeghi, A., Khataee, A., Abedi, M., Tarrahi, R., Dadpour, M., & Vafaei, F. (2018). Effects of TiO_2 nanoparticles on the aquatic plant *Spirodela polyrrhiza*: Evaluation of growth parameters, pigment contents, and antioxidant enzyme activities. *Journal of Environmental Sciences (China), 64*, 130–138. https://doi.org/10.1016/j.jes.2016.12.020.

Navarro, E., Baun, A., Behra, R., Hartmann, N. B., Filser, J., Miao, A. J., Quigg, A., Santschi, P. H., & Sigg, L. (2008). Environmental behavior and ecotoxicity of engineered nanoparticles to algae, plants, and fungi. *Ecotoxicology, 17*(5), 372–386. https://doi.org/10.1007/s10646-008-0214-0.

Nayan, R., Rawat, M., Negi, B., Pande, A., & Arora, S. (2016). Zinc sulfide nanoparticle-mediated alterations in growth and antioxidant status of *Brassica juncea*. *Biologia, 71*(8), 896–902. https://doi.org/10.1515/biolog-2016-0107.

Nguyen, D. T. C., Le, H. T. N., Nguyen, T. T., Nguyen, T. T. T., Bach, L. G., Nguyen, T. D., & Tran, T. V. (2021). Multifunctional ZnO nanoparticles bio-fabricated from *Canna indica* L. flowers for seed germination, adsorption, and photocatalytic degradation of organic dyes. *Journal of Hazardous Materials, 420*, 126586. https://doi.org/10.1016/j.jhazmat.2021.126586.

Obroucheva, N. V. (2013). Aquaporins in seeds. *Seed Science Research, 23*(4), 213–216. https://doi.org/10.1017/S0960258513000275.

Parisi, C., Vigani, M., & Rodriguez, C. E. (2014). Proceedings of a workshop on "Nanotechnology for the agricultural sector: From research to the field." *JRC Publications Repository*. https://doi.org/10.2791/80497.

Pérez-de-Luque, A. (2017). Interaction of Nanomaterials with Plants: What Do We Need for Real Applications in Agriculture? *Frontiers in Environmental Science, 5*. https://www.frontiersin.org/articles/10.3389/fenvs.2017.00012.

Pradhan, S., Patra, P., Das, S., Chandra, S., Mitra, S., Dey, K. K., Akbar, S., Palit, P., & Goswami, A. (2013). Photochemical Modulation of Biosafe Manganese Nanoparticles on *Vigna radiata*: A Detailed Molecular, Biochemical, and Biophysical Study. *Environmental Science & Technology, 47*(22), 13122–13131. https://doi.org/10.1021/es402659t.

Queiroz, A. M., Mezacasa, A. V., Graciano, D. E., Falco, W. F., M'Peko, J. C., Guimarães, F. E. G., Lawson, T., Colbeck, I., Oliveira, S. L., & Caires, A. R. L. (2016). Quenching of chlorophyll fluorescence induced by silver nanoparticles. *Spectrochimica Acta. Part*

A, Molecular and Biomolecular Spectroscopy, *168*, 73–77. https://doi.org/10.1016/j.saa.2016.05.033.

Radchuk, V., & Borisjuk, L. (2014). Physical, metabolic and developmental functions of the seed coat. *Frontiers in Plant Science*, *5*. https://www.frontiersin.org/articles/10.3389/fpls.2014.00510.

Rai-Kalal, P., & Jajoo, A. (2021). Priming with zinc oxide nanoparticles improve germination and photosynthetic performance in wheat. *Plant Physiology and Biochemistry*, *160*, 341–351. https://doi.org/10.1016/J.PLAPHY.2021.01.032.

Raliya, R., & Tarafdar, J. C. (2013). ZnO nanoparticle biosynthesis and its effect on phosphorous-mobilizing enzyme secretion and gum contents in clusterbean (*Cyamopsis tetragonoloba* L.). *Agricultural Research*, *2*(1), 48–57. https://doi.org/10.1007/s40003-012-0049-z.

Raliya, R., Franke, C., Chavalmane, S., Nair, R., Reed, N., & Biswas, P. (2016). Quantitative understanding of nanoparticle uptake in watermelon plants. *Frontiers in Plant Science*, *7*. https://www.frontiersin.org/articles/10.3389/fpls.2016.01288.

Rani, A., Rani, K., Tokas, J., Anamika, Singh, A., Kumar, R., Punia, H., & Kumar, S. (2020). Nanomaterials for Agriculture Input Use Efficiency. In S. Kumar, R. S. Meena, & M. K. Jhariya (Eds.), *Resources Use Efficiency in Agriculture* (pp. 137–175). Springer. https://doi.org/10.1007/978-981-15-6953-1_5.

Rani, P., Kaur, G., Rao, K. V., Singh, J., & Rawat, M. (2020). Impact of green synthesized metal oxide nanoparticles on seed germination and seedling growth of *Vigna radiata* (mungbean) and *Cajanus cajan* (red gram). *Journal of Inorganic and Organometallic Polymers and Materials*, *30*(10), 4053–4062. https://doi.org/10.1007/s10904-020-01551-4.

Rasool, N. (2022). Plant Hormones: Role in Alleviating Biotic Stress. In *Plant Hormones—Recent Advances, New Perspectives and Applications*. IntechOpen. https://doi.org/10.5772/intechopen.102689.

Rawat, P. S., Kumar, R., Ram, P., & Pandey, P. (2018). Effect of nanoparticles on wheat seed germination and seedling growth. *International Journal of Agricultural and Biosystems Engineering*, *12*(1), 13–16.

Reddy Pullagurala, V. L., Adisa, I. O., Rawat, S., Kalagara, S., Hernandez-Viezcas, J. A., Peralta-Videa, J. R., & Gardea-Torresdey, J. L. (2018). ZnO nanoparticles increase photosynthetic pigments and decrease lipid peroxidation in soil grown cilantro (*Coriandrum sativum*). *Plant Physiology and Biochemistry*, *132*, 120–127. https://doi.org/10.1016/J.PLAPHY.2018.08.037.

Rico, C. M., Majumdar, S., Duarte-Gardea, M., Peralta-Videa, J. R., & Gardea-Torresdey, J. L. (2011). Interaction of nanoparticles with edible plants and their possible implications in the food chain. *Journal of Agricultural and Food Chemistry*, *59*(8), 3485–3498. https://doi.org/10.1021/jf104517j.

Rostami, M., Talarposhti, R. M., Mohammadi, H., & Demyan, M. S. (2019). Morpho-physiological Response of Saffron (*Crocus sativus* L.) to Particle Size and Rates of Zinc Fertilizer, *50*(10), 1250–1257. https://doi.org/10.1080/00103624.2019.1614602.

Saquib, Q., Faisal, M., Alatar, A. A., Al-Khedhairy, A. A., Ahmed, M., Ansari, S. M., Alwathnani, H. A., Okla, M. K., Dwivedi, S., Musarrat, J., Praveen, S., Khan, S. T., Wahab, R., Siddiqui, M. A., & Ahmad, J. (2016). Genotoxicity of ferric oxide nanoparticles in *Raphanus sativus*: Deciphering the role of signaling factors, oxidative stress and cell death. *Journal of Environmental Sciences, 47*, 49–62. https://doi.org/10.1016/j.jes.2015.12.037.

Sarmast, M. K., & Salehi, H. (2016). Silver Nanoparticles: An Influential Element in Plant Nanobiotechnology. *Molecular Biotechnology, 58*(7), 441–449. https://doi.org/10.1007/s12033-016-9943-0.

Sattelmacher, B. (2001). The apoplast and its significance for plant mineral nutrition. *New Phytologist, 149*(2), 167–192. https://doi.org/10.1046/j.1469-8137.2001.00034.x.

Schlesinger, W. (2020). The carbon cycle of terrestrial ecosystems (pp. 141–182). https://doi.org/10.1016/B978-0-12-814608-8.00005-0.

Schwab, F., Zhai, G., Kern, M., Turner, A., Schnoor, J. L., & Wiesner, M. R. (2016). Barriers, pathways and processes for uptake, translocation and accumulation of nanomaterials in plants–Critical review. *Nanotoxicology, 10*(3), 257–278. https://doi.org/10.3109/17435390.2015.1048326.

Segura, R., Vásquez, G., Colson, E., Gerbaux, P., Frischmon, C., Nesic, A., García, D. E., & Cabrera-Barjas, G. (2020). Phytostimulant properties of highly stable silver nanoparticles obtained with saponin extract from *Chenopodium quinoa*. *Journal of the Science of Food and Agriculture, 100*(13), 4987–4994. https://doi.org/10.1002/jsfa.10529.

Shabbir, A., Khan, M. M. A., Ahmad, B., Sadiq, Y., Jaleel, H., & Uddin, M. (2019). Efficacy of TiO_2 nanoparticles in enhancing the photosynthesis, essential oil and khusimol biosynthesis in *Vetiveria zizanoides* L. Nash. *Photosynthetica, 57*(2), 599–606. https://doi.org/10.32615/ps.2019.071.

Sharma, P., Bhatt, D., Zaidi, M. G. H., Saradhi, P. P., Khanna, P. K., & Arora, S. (2012). Silver Nanoparticle-Mediated Enhancement in Growth and Antioxidant Status of *Brassica juncea*. *Applied Biochemistry and Biotechnology, 167*(8), 2225–2233. https://doi.org/10.1007/s12010-012-9759-8.

Shekhar, S., Sharma, S., Kumar, A., Taneja, A., & Sharma, B. (2021). The framework of nanopesticides: A paradigm in biodiversity. *Materials Advances, 2*(20), 6569–6588. https://doi.org/10.1039/D1MA00329A.

Shen, C. X., Zhang, Q. F., Li, J., Bi, F., & Yao, N. (2010). Induction of programmed cell death in Arabidopsis and rice by single-wall carbon nanotubes. *American Journal of Botany, 97*, 1602–1609. https://doi.org/10.3732/ajb.1000073.

Shweta, ., Tripathi, D. K., Singh, S., Singh, S., Dubey, N. K., & Chauhan, D. K. (2016). Impact of nanoparticles on photosynthesis: Challenges and opportunities. *Materials Focus, 5*(5), 405–411. https://doi.org/10.1166/mat.2016.1327.

Singh, A., Singh, N. B., Hussain, I., Singh, H., Yadav, V., & Singh, S. C. (2016). Green synthesis of nano zinc oxide and evaluation of its impact on germination and metabolic activity of *Solanum lycopersicum*. *Journal of Biotechnology, 233*, 84–94. https://doi.org/10.1016/j.jbiotec.2016.07.010.

Singh, J., Kumar, S., Alok, A., Upadhyay, S. K., Rawat, M., Tsang, D. C. W., Bolan, N., & Kim, K. H. (2019). The potential of green synthesized zinc oxide nanoparticles as a nutrient source for plant growth. *Journal of Cleaner Production, 214*, 1061–1070. https://doi.org/10.1016/j.jclepro.2019.01.018.

Singh, N., & Sarma, S. (2022). Biological routes of hydrogen production: A critical assessment. In S. Sahay (Ed.), *Handbook of Biofuels* (pp. 419–434). Academic Press. https://doi.org/10.1016/B978-0-12-822810-4.00021-X.

Singh, Y., Kaushal, S., & Sodhi, R. S. (2020). Biogenic synthesis of silver nanoparticles using cyanobacterium *Leptolyngbya* sp. WUC 59 cell-free extract and their effects on bacterial growth and seed germination. *Nanoscale Advances, 2*(9), 3972–3982. https://doi.org/10.1039/D0NA00357C.

Song, G., Gao, Y., Wu, H., Hou, W., Zhang, C., & Ma, H. (2012). Physiological effect of anatase TiO_2 nanoparticles on *Lemna minor*. *Environmental Toxicology and Chemistry, 31*(9), 2147–2152. https://doi.org/10.1002/etc.1933.

Song, U., Jun, H., Waldman, B., Roh, J., Kim, Y., Yi, J., & Lee, E. J. (2013). Functional analyses of nanoparticle toxicity: A comparative study of the effects of TiO_2 and Ag on tomatoes (*Lycopersicon esculentum*). *Ecotoxicology and Environmental Safety, 93*, 60–67. https://doi.org/10.1016/j.ecoenv.2013.03.033.

Sun, D., Hussain, H. I., Yi, Z., Siegele, R., Cresswell, T., Kong, L., & Cahill, D. M. (2014). Uptake and cellular distribution, in four plant species, of fluorescently labeled mesoporous silica nanoparticles. *Plant Cell Reports, 33*(8), 1389–1402. https://doi.org/10.1007/s00299-014-1624-5.

Sun, Y., Jing, R., Zheng, F., Zhang, S., Jiao, W., & Wang, F. (2019). Evaluating phytotoxicity of bare and starch-stabilized zero-valent iron nanoparticles in mung bean. *Chemosphere, 236*, 124336. https://doi.org/10.1016/j.chemosphere.2019.07.067.

Syu, Y., Hung, J. H., Chen, J. C., & Chuang, H. (2014). Impacts of size and shape of silver nanoparticles on Arabidopsis plant growth and gene expression. *Plant Physiology and Biochemistry, 83*, 57–64. https://doi.org/10.1016/j.plaphy.2014.07.010.

Taha, R. A., Hassan, M. M., Ibrahim, E. A., Abou Baker, N. H., & Shaaban, E. A. (2016). Carbon nanotubes impact on date palm *in vitro* cultures. *Plant Cell, Tissue and Organ Culture, 127*(2), 525–534. https://doi.org/10.1007/s11240-016-1058-6.

Tan, W., Du, W., Barrios, A. C., Armendariz, R., Zuverza-Mena, N., Ji, Z., Chang, C. H., Zink, J. I., Hernandez-Viezcas, J. A., Peralta-Videa, J. R., & Gardea-Torresdey, J. L. (2017). Surface coating changes the physiological and biochemical impacts of nano-TiO_2 in basil (*Ocimum basilicum*) plants. *Environmental Pollution, 222*, 64–72. https://doi.org/10.1016/j.envpol.2017.01.002.

Teszlák, P., Kocsis, M., Scarpellini, A., Jakab, G., & Kőrösi, L. (2018). Foliar exposure of grapevine (*Vitis vinifera* L.) to TiO_2 nanoparticles under field conditions: Photosynthetic response and flavonol profile. *Plant Cell Reports*. https://doi.org/10.1007/s11099-018-0832-6.

Tripathi, D. K., Shweta., Singh, S., Singh, S., Pandey, R., Singh, V. P., Sharma, N. C., Prasad, S. M., Dubey, N. K., & Chauhan, D. K. (2017). An overview on manufactured nanoparticles in plants: Uptake, translocation, accumulation and phytotoxicity. *Plant Physiology and Biochemistry, 110*, 2–12. https://doi.org/10.1016/j.plaphy.2016.07.030.

Tripathi, D. K., Singh, S., Singh, S., Srivastava, P. K., Singh, V. P., Singh, S., Prasad, S. M., Singh, P. K., Dubey, N. K., Pandey, A. C., & Chauhan, D. K. (2017). Nitric oxide alleviates silver nanoparticles (AgNps)-induced phytotoxicity in *Pisum sativum* seedlings. *Plant Physiology and Biochemistry, 110*, 167–177. https://doi.org/10.1016/j.plaphy.2016.06.015.

Tripathi, D. K., Singh, V. P., Gangwar, S., Prasad, S. M., Maurya, J. N., & Chauhan, D. K. (2013). Role of silicon in the enrichment of plant nutrients and protection from biotic and abiotic stresses. *Improvement of Crops in the Era of Climatic Changes, 39*.

Tripathi, D., Singh, M., & Pandey-Rai, S. (2022). Crosstalk of nanoparticles and phytohormones regulate plant growth and metabolism under abiotic and biotic stress. *Plant Stress, 6*, 100107. https://doi.org/10.1016/j.stress.2022.100107.

Tumburu, L., Andersen, C. P., Rygiewicz, P. T., & Reichman, J. R. (2015). Phenotypic and genomic responses to titanium dioxide and cerium oxide nanoparticles in Arabidopsis germinants. *Environmental Toxicology and Chemistry, 34*(1), 70–83. https://doi.org/10.1002/etc.2756.

Vannini, C., Domingo, G., Onelli, E., De Mattia, F., Bruni, I., Marsoni, M., & Bracale, M. (2014). Phytotoxic and genotoxic effects of silver nanoparticles exposure on germinating wheat seedlings. *Journal of Plant Physiology, 171*(13), 1142–1148. https://doi.org/10.1016/j.jplph.2014.05.002.

Vannini, C., Domingo, G., Onelli, E., Prinsi, B., Marsoni, M., Espen, L., & Bracale, M. (2013). Morphological and proteomic responses of *Eruca sativa* exposed to silver nanoparticles or silver nitrate. *PLOS ONE, 8*(7), e68752. https://doi.org/10.1371/journal.pone.0068752.

Wahid, I., Kumari, S., Ahmad, R., Hussain, S. J., Alamri, S., Siddiqui, M. H., & Khan, M. I. R. (2020). Silver nanoparticle regulates salt tolerance in wheat through changes in ABA concentration, ion homeostasis, and defense systems. *Biomolecules, 10*(11), 11. https://doi.org/10.3390/biom10111506.

Wang, H., Zhang, M., Song, Y., Li, H., Huang, H., Shao, M., Liu, Y., & Kang, Z. (2018). Carbon dots promote the growth and photosynthesis of mung bean sprouts. *Carbon, 136*, 94–102. https://doi.org/10.1016/j.carbon.2018.04.051.

Wang, X. P., Li, Q. Q., Pei, Z. M., & Wang, S. C. (2018). Effects of zinc oxide nanoparticles on the growth, photosynthetic traits, and antioxidative enzymes in tomato plants. *Biologia Plantarum, 62*(4), 801–808. https://doi.org/10.1007/s10535-018-0813-4.

Wang, Z., Xie, X., Zhao, J., Liu, X., Feng, W., White, J. C., & Xing, B. (2012). Xylem- and phloem-based transport of CuO nanoparticles in maize (*Zea mays* L.). *Environmental Science & Technology, 46*(8), 4434–4441. https://doi.org/10.1021/es204212z.

Wei, J., Zou, Y., Li, P., & Yuan, X. (2020). Titanium dioxide nanoparticles promote root growth by interfering with auxin pathways in *Arabidopsis thaliana. Phyton-International Journal of Experimental Botany, 89*(4). https://doi.org/10.32604/phyton.2020.010973.

Wong, M. H., Misra, R. P., Giraldo, J. P., Kwak, S. Y., Son, Y., Landry, M. P., Swan, J. W., Blankschtein, D., & Strano, M. S. (2016). Lipid exchange envelope penetration (LEEP) of nanoparticles for plant engineering: A universal localization mechanism. *Nano Letters, 16*(2), 1161–1172. https://doi.org/10.1021/acs.nanolett.5b04467.

Xin, X., Zhao, F., Judy, J. D., & He, Z. (2022). Copper stress alleviation in corn (*Zea mays* L.): Comparative efficiency of carbon nanotubes and carbon nanoparticles. *NanoImpact, 25*, 100381. https://doi.org/10.1016/j.impact.2022.100381.

Yadu, B., Chandrakar, V., Korram, J., Satnami, M. L., Kumar, M., & S, K. (2018). Silver nanoparticle modulates gene expressions, glyoxalase system and oxidative stress markers in fluoride stressed *Cajanus cajan* L. *Journal of Hazardous Materials, 353*, 44–52. https://doi.org/10.1016/j.jhazmat.2018.03.061.

Yoon, H., Kang, Y. G., Chang, Y. S., & Kim, J. H. (2019). Effects of zerovalent iron nanoparticles on photosynthesis and biochemical adaptation of soil-grown *Arabidopsis thaliana. Nanomaterials, 9*(11). https://doi.org/10.3390/nano9111543.

Yue, L., Ma, C., Zhan, X., White, J. C., & Xing, B. (2017). Molecular mechanisms of maize seedling response to La2O3 NP exposure: Water uptake, aquaporin gene expression and signal transduction. *Environmental Science: Nano, 4*(4), 843–855. https://doi.org/10.1039/C6EN00487C.

Zahedi, S. M., Abdelrahman, M., Hosseini, M. S., Hoveizeh, N. F., & Tran, L. S. P. (2019). Alleviation of the effect of salinity on growth and yield of strawberry by foliar spray of selenium-nanoparticles. *Environmental Pollution, 253*, 246–258. https://doi.org/10.1016/j.envpol.2019.04.078.

Zamir, D. (2001). Improving plant breeding with exotic genetic libraries. *Nature Reviews Genetics, 2*(12), Article 12. https://doi.org/10.1038/35103590.

Zhang, H., Yue, M., Zheng, X., Xie, C., Zhou, H., & Li, L. (2017). Physiological effects of single- and multi-walled carbon nanotubes on rice seedlings. *IEEE Transactions on Nanobioscience, 16*(7), 563–570. https://doi.org/10.1109/TNB.2017.2715359.

Zhang, Y., Liu, N., Wang, W., Sun, J., & Zhu, L. (2020). Photosynthesis and related metabolic mechanism of promoted rice (*Oryza sativa* L.) growth by TiO_2 nanoparticles. *Frontiers of Environmental Science & Engineering, 14*(6), 1–12. https://doi.org/10.1007/S11783-020-1282-5.

Zhang, Z., Ke, M., Qu, Q., Peijnenburg, W. J. G. M., Lu, T., Zhang, Q., Ye, Y., Xu, P., Du, B., Sun, L., & Qian, H. (2018). Impact of copper nanoparticles and ionic copper exposure on wheat (*Triticum aestivum* L.) root morphology and antioxidant response. *Environmental Pollution, 239*, 689–697. https://doi.org/10.1016/j.envpol.2018.04.066.

Zhao, L., Hu, Q., Huang, Y., & Keller, A. A. (2017). Response at genetic, metabolic, and physiological levels of maize (*Zea mays*) exposed to a $Cu(OH)_2$ nanopesticide. *ACS Sustainable Chemistry & Engineering, 5*(9), 8294–8301. https://doi.org/10.1021/acssuschemeng.7b01968.

Zhao, L., Huang, Y., Adeleye, A. S., & Keller, A. A. (2017). Metabolomics reveals $Cu(OH)_2$ nanopesticide-activated anti-oxidative pathways and decreased beneficial antioxidants in spinach leaves. *Environmental Science & Technology, 51*(17), 10184–10194. https://doi.org/10.1021/acs.est.7b02163.

Zhu, Z. J., Wang, H., Yan, B., Zheng, H., Jiang, Y., Miranda, O. R., Rotello, V. M., Xing, B., & Vachet, R. W. (2012). Effect of surface charge on the uptake and distribution of gold nanoparticles in four plant species. *Environmental Science & Technology, 46*(22), 12391–12398. https://doi.org/10.1021/es301977w.

Zuverza-Mena, N., Armendariz, R., Peralta-Videa, J. R., & Gardea-Torresdey, J. L. (2016). Effects of silver nanoparticles on radish sprouts: Root growth reduction and modifications in the nutritional value. *Frontiers in Plant Science, 7*, 90. https://doi.org/10.3389/fpls.2016.00090.

CHAPTER 8

Economical and Market Status of Nanopesticides

VINAY ASERI,[1] POONAM KUMARI,[1] VANISHA GODARA,[1] GARIMA AWASTHI,[2] TARUN KUMAR KUMAWAT,[3] SHALINI FAGNA,[3] G. P. SINGH,[3] and HARSH PANDEY[4]

[1]*Department of Forensic Science, Vivekananda Global University, Jaipur, Rajasthan, India*

[2]*K. R. Mangalam University, Gurugram, Haryana, India*

[3]*Department of Botany, University of Rajasthan, Jaipur, Rajasthan, India*

[4]*Department of Biotechnology and Chemical Engineering, School of Civil and Chemical Engineering, Manipal University, Jaipur, Rajasthan, India*

ABSTRACT

In order to improve agriculture production and feed the growing global population, agriculture and its allied industries face significant hurdles. The excessive use of pesticides has a number of negative effects, including contamination of the environment, global warming, the emergence of pests that are resistant to treatment, environmental imbalance, toxicity to humans and other organisms, and so on. To ensure the sustainable growth of agriculture in this situation, precision agricultural techniques are required. Consequently, nanotechnology may offer effective answers for the existing issues in the agricultural sector. The economic viability of the technology for the expansion of sustainable agriculture is highlighted in this chapter's assessment of nanotechnology techniques for agricultural

Nanopesticides: A Perspective from Welfare to Risk Assessment.
Garima Awasthi, Payal Lodha, and Gajendra Pal Singh (Eds.)

DOI: 10.1201/9781003605874-8

activities. The market size of the pesticides market was estimated to be US$105.05 billion in 2021, and it is predicted that from 2022 to 2029, total revenue would increase by 11.5%, with the pesticides market likely to reach close to US$250.96 billion. The disadvantages and difficulties that prevent the commercialization of nanopesticides were also assessed. The use of nanotechnology in agriculture has recently emerged as a cutting-edge, quickly expanding research area. The use of nanotechnology advancements in the field of nanopesticides has enabled the transition of materials used in nanopesticides from macro to micro sizes. The market for nanopesticides was estimated at $0.6 billion in 2021 and projected to reach $1.9 billion by 2030, with a CAGR of 13.66% between 2022 and 2030. As a result, nanopesticides are increasingly having a significant impact on global economic status and agricultural production.

8.1 INTRODUCTION

This chapter focuses on the market demand and evaluation application of nanotechnology to produce nanofertilizers or nanopesticides, often known as "nanoagrochemicals," for plant nourishment and protection. Modern agriculture largely depends on the use of agrochemicals. Nevertheless, in comparison to other components in the food chain, including food production or packing, nanopesticides and nanofertilizers have received lesser attention. Regarding possible environmental effects, nanoagrochemicals may be viewed as being very significant since these chemicals are directly and purposefully applied to the environment, and they would be the environment's key targeted ubiquitous source of produced nanoparticles. (Kah et al., 2013).

There are certain concerns to human and environmental health linked with the use of agrochemicals (for instance, residues on food items and the water resources pollution). According to several sources, the emergence of high-tech agricultural sectors facilitated by nanotechnology will be possible. Inputs like herbicides, fertilizers, and water will be precisely controlled and monitored in these fields with the help of a wide range of intelligent nanotools (Scott and Chen, 2002; and so forth, 2004; Sekhon, 2014; Liu and Lal, 2015). The creation of such tools will undoubtedly revolutionize agricultural methods, and they may even help to significantly reduce the environmental effect of current agriculture while also increasing crop quantity and quality. Nevertheless, it must be understood that such approaches are inappropriate in the current economic scenario and also pose a chance of high risk from initial stage regulation and public criticism due to the poor profit margins in agriculture. In general, agrochemicals do not support the notion that considerably smaller equals more sensitive and hence more effective. Most "nano" pesticides that have been identified in literature are far bigger than the recommended size limit of 100 nm for regulatory considerations (discussed in Kah et al., 2013). There are several questions around the nature of nanoparticles and how nanopesticides can be impacted by the proposed standards. The far more important point is that many new so-called nanoformulations would be excluded by a definition that just considers size while including items that have been available on the market for decades without causing any specific issues (such as emulsifiers, formulants like polymers and clays, and so forth). Therefore, the EU's plan to create a database for nanomaterials, carries a possibility of confusing consumers even more by adding components which are in use for years without being labelled as "nano" (EC, 2014). Instead of concentrating just on the nanoparticles and how they should be specified, it could be more appropriate to discuss about formulations or nanotechnology (JRC-IPTS, 2014). However, depending on the context, the phrases "nanopesticide" and "nanofertilizer" might have quite different meanings. A huge variety of products can be designated by the terms depending on their nature, size, relevance for agricultural practices and level of development, according to inventories that have been introduced so far, with a focus on scientific literature and patent evaluation. The prefix "nano" has been implicitly identified in scientific literature with the idea of innovation and superior qualities compared to non-nanoequivalents. In order to garner more attention and maybe make publishing easier, numerous formulations were given the name "nano" (e.g., Gogos et al., 2012; Kah et al., 2013). In order to review the items that may soon be available on the market and their suitability with the current legal and business frameworks, a number of critical topics are addressed after briefly describing the nanoagrochemical-related activities carried out during the past 10 years (Figure 8.1). Discussion of the potential future of nanoagrochemicals integrates perspectives from the scientific, industrial, and regulatory sectors. Finally, new options that could enable the agrochemical industry to benefit from nanotechnology and perhaps lessen its effects on human and environmental health are offered.

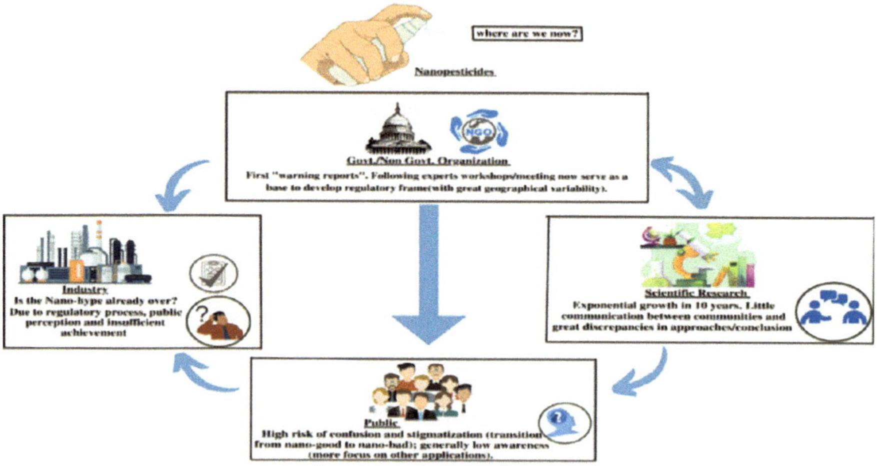

FIGURE 8.1 Global market economic viability of nanoagrochemicals.

8.2 THE MARKET SIZE OF NANOPESTICIDE

The potential stigmatization of nanomaterials is a significant concern. The combination of food, pesticides, and nanotechnology, in particular, has a high potential to raise public alarm, as noted by the European Crop Protection Association. The crop protection industry worries about a scenario resembling the public rejection of genetically modified organisms (GMOs) (JRC-IPTS, 2014). Despite various public awareness campaigns, the general public appears to have limited knowledge of nanoagrochemicals, even as nanoparticles are now intentionally integrated at multiple stages within the food chain. This limited awareness may be due to (i) a primary focus on nanoparticles used directly in food and food packaging as ingredients or additives (such as titanium dioxide (TiO_2) and silver nanoparticles (nanoAg)) and (ii) fundamental concerns about the use of agrochemicals in general, as pesticides do not need to be "nano" to raise objections (Friends of the Earth, 2008).

Until a specific definition is established for nanoagrochemicals, conflicting claims across societies and sectors are likely to continue. Some government authorities argue that nano-enabled pesticides remain a future possibility, stating that "no authorized pesticides on the market incorporate nanomaterials" (S. Ramaswamy, Director of the USDA National Institute of Food and Agriculture, cited in Gewin, 2015; EFSA, 2009).

Conversely, officials from some businesses and research institutions have suggested that "extremely few, if any," or "not many" exist (JRC-IPTS, 2014). Other organizations have issued warnings that "manufactured nanoparticles, nano-emulsions, and nano-capsules are now found in agricultural chemicals" and that "the first nano-formulations of pesticides are silently making their way onto agricultural fields" (Gewin, 2015).

8.3 MARKET USABLE TYPES OF NANOPESTICIDES

The nanoformulation of traditional pesticides using polymers or metal nanoparticles is one of the most challenging areas in the pesticide industry. According to Pérez-de-Luque and Rubiales (2009) and Agrawal and Rathore (2014), nanoencapsulation of pesticides offers advantages by restricting and slowing the release of the active ingredient. By modifying the nanocapsule's outer shell, these formulations can release low doses of pesticides over extended periods, reducing excess discharge and minimizing unwanted pesticide release. Another benefit of nanocarriers in plant protection is site-targeted distribution and enhanced stability of the active component (Nair et al., 2010). Similarly, nanoemulsions—created with water or oil—can improve a pesticide's solubility and increase its effectiveness against various pests (Liu et al., 2006; Wang et al., 2007). Table 8.2 summarizes the different types of nanopesticides developed.

Several phytopathogens, including *Bipolaris sorokiniana*, *Colletotrichum gloeosporioides*, *Botrytis cinerea*, *Pythium ultimum*, *Phoma*, *Magnaporthe grisea*, *Trichoderma* sp., *Sclerotinia sclerotiorum*, *Sphaerotheca pannosa*, *Fusarium culmorum*, and *Rhizoctonia solani*, have been found susceptible to silver nanoparticles (AgNPs) (Gajbhiye et al., 2009; Park, 2006; Gopal et al., 2011; Kumar et al., 2010). Additionally, silver-containing silica nanoparticles (Si-Ag NPs) have shown 100% effectiveness in treating cucumbers affected by powdery mildew. Porous silica nanomaterials have also been reported as validamycin carriers, providing controlled pesticide delivery (Liu et al., 2006; Park et al., 2006). Copper nanoparticles (CuNPs) at low concentrations have demonstrated resistance against *Xanthomonas oryzae*, *Phytophthora infestans*, *Xanthomonas campestris*, as well as several fungal pathogens and Gram-positive and Gram-negative bacteria (see Table 8.1 and Figure 8.1) (Cioffi et al., 2005; Esteban-Tejeda et al., 2009; Giannousi et al., 2014).

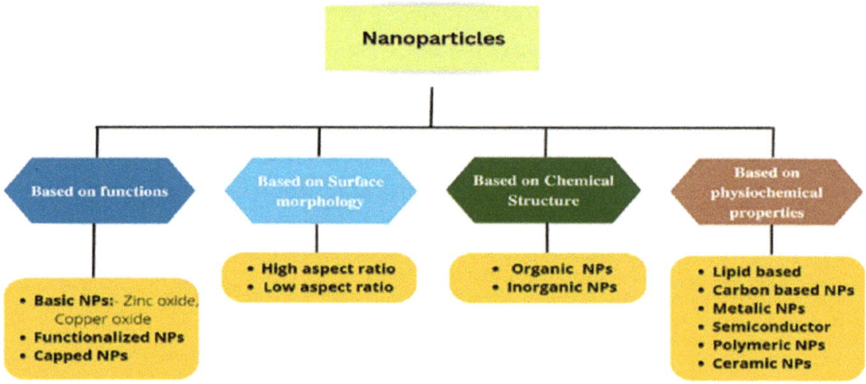

FIGURE 8.2 Demanding nanoparticles use in economical agriculture market.

Penicillium expansum, A. Niger B. cinerea, Aspergillus flavus, and *Penicillium expansum,* were all resistant to the antifungal effects of Zn NP, which has been described as a nanofertilizer (He et al., 2011). ZnO based nanopesticide is a preferred pesticide alternative since it is less hazardous to plants than Ag NP (Jayaseelan et al., 2012). Similarly, nanosulfur, nanohexaconazole is also very efficient against the red spider mite *Tetranychusutricae*, *Erysiphe cichoracearum*, and *R. solani*, which are plant harmful fungi. Prior to their commercialization, it is critical to evaluate these nanomaterials' effects on the ecosystem and compliance with the Toxic Substance Control Act. Federal authorities must control the Nanopesticide Act to avoid their harmful effects on the environment.

8.4 WORLD MARKET: NANOPESTICIDES AVAILABILITY AND LIMITATIONS

Pesticides are often used in commercial agriculture, and new, effective, and target-specific pesticides are constantly being produced for this reason. A significant number of pesticides are assessed each year due to the estimated 2 million tonnes of pesticides that are used annually nationwide. Additionally, it has been shown that only 0.1% of the pesticides applied really reach the intended pests, with the other 99.9% remaining in the environment and having adverse effects on both human health and the environment. It is interesting that, in contrast to each life lost as a result of pesticides, there would be a factor of 1000 more deaths if pesticides were not used

TABLE 8.1 List of Several Types Nanopesticides Use in Market and Functions

Sl. No.	Engineered Nanoparticles	Synthesis Method	Effective Against	Effects	Reference
1	Copper nanoparticles	Chemical method	*Alternaria alternata*, *Phoma destructiva*, and *Curvularia lunata*	Novel antifungal action against plant pathogenic fungi; effective as a disinfectant in poultry and animal agriculture	Wilson et al. (2014)
2	Cu_2O	Solvothermal approach	*Saccharomyces cerevisiae*	Significant antifungal activity; impacts lipid peroxidation and reactive oxygen species generation	Wilson et al. (2014)
3	Cu-chitosan nanoparticles	Ionic gelation method	*Macrophomina phaseolina*, *Rhizoctonia solani*, and *Alternaria alternata*	At low concentrations (0.1%), excellent fungal growth suppression (89.5%, 63.0%, and 60.1% growth inhibition, respectively)	Gopal et al. (2011)
4	Chitosan and laminarian	—	*Plasmopara viticola*	Effective in reducing amino acids and controlling grapevine downy mildew	Gopal et al. (2011)
5	ZnO-nanocopper-loaded silica gel	Chemical method	Citrus canker disease	Effectively prevents citrus canker in 'Ruby' grapefruit trees; outperforms Cu_2O and ZnO insecticides at less than half the metallic rate	Rema and Martinez (2013)
6	Ag-doped hollow TiO_2 nanoparticles	—	*Venturia inaequalis* and *Fusarium solani*	Enhanced fungicidal action under visible light	Nair et al. (2010)
7	Silver nanoparticles	Biosynthesis using various bacterial strains	Bean yellow mosaic virus	Antiviral agent for fava bean crops	Saharan et al. (2013)
8	Silver nanoparticles	Biosynthesis using *Acalypha indica* leaf extracts	*Escherichia coli* and *Vibrio cholerae*	Strong antimicrobial activity against waterborne bacterial pathogens	Rema and Martinez (2013)
9	Silver nanoparticles	Fungal-mediated biosynthesis by *Trichoderma harzianum*	White mold (*Sclerotinia sclerotiorum*)	No adverse effects on soybean growth; cytotoxic and genotoxic effects on *S. sclerotiorum* germination	Nair et al. (2010)

TABLE 8.1 *(Continued)*

Sl. No.	Engineered Nanoparticles	Synthesis Method	Effective Against	Effects	Reference
10	Silver and gold nanoparticles	Biosynthesis using *Chrysosporium tropicum*	*Aedes aegypti*	Gold nanoparticles show approximately three times higher mortality in *Aedes aegypti* compared to silver nanoparticles	Saharan et al. (2013)
11	Nanostructured alumina	Combustion synthesis technique	*Rhyzopertha dominica* (F.) and *Sitophilus oryzae* (L.)	Comparable mortality rates to commercial insecticidal dusts (500–5000 mg/kg)	Ho et al. (2015)
12	Mesoporous alumina nanoparticles	Microemulsion method	*Fusarium oxysporum*	Strong fungal growth suppression (78%) in tomato crops at 400 mg/L	Ho et al. (2015)
13	Catalysis reaction technique	Catalysis reaction technique	*Fusarium solani* and *Venturia inaequalis*	Superior fungicidal activity in smaller nanoparticles (35 nm) for diseased tomato leaves and apple scab, compared to larger nanoparticles (87 and 200 nm)	Rajesh et al. (2012)
14	PVP-Ag and Ag^+ nanoparticles	Chemical method	Arbuscular mycorrhizal colonization	Excellent reduction in bacterial, actinomycete, and fungal populations compared to Ag_2S for PVP-Ag and Ag^+ nanoparticles	Rajesh et al. (2012)

TABLE 8.2 Different Types Nanopesticides Target on Plants Pathogen Crop Disease

SL. No.	Nanomaterial	Crop Disease	Plant Pathogen	Impact	References
1	Silver (Ag) NP	Bacterial blight	*Xanthomonas campestris* pv. *campestris*	Substantial bacterial reduction	Rajesh et al. (2012)
2	Ag (graphene oxide composite)	Bacterial spot disease	*Xanthomonas*	Considerable bacterial growth inhibition	Ocsoy et al. (2013)
3	Ag core-DHPAC shell nanocluster	Fungal disease	*Phytophthora nicotianae*, *Phytophthora colocasiae*, and *Phytophthora capsici*	Growth inhibition up to 80%	Ho et al. (2015)
4	Chitosan NP	Disease in chili (Fungal disease)	*Colletotrichum capsici*, *Colletotrichum gloeosporioides*, *Aspergillus niger*, *Rhizopus* sp., *Macrophomina phaseolina*, *Rhizoctonia solani*, and *Alternaria alternata*	Slow mycelial development; spore germination is prevented	Chookhongkha et al. (2012); Saharan et al. (2013)
5	Iron oxide (FeO) NP	Fungal disease	*Botrytis cinerea*, *Fusarium oxysporum*, and *Rhizoctonia solani*	60–80% fungal inhibition	Chhipa and Kaushik (2015)
6	Copper (Cu) NP	Fungal disease in pomegranate, bacterial blight, and disease in tomato	*Fusarium* sp., *Xanthomonas axonopodis* pv. *punicae*, and *Phytophthora infestans*	Exhibits antifungal properties; inhibits bacterial growth; active as an antifungal agent at low concentration	Bramhanwade et al. (2016); Mondal and Mani (2012); Giannousi et al. (2013)

in agricultural practices throughout the world. The size of the pesticides market was estimated to be US$105.05 billion in 2021, and it is predicted that from 2022 to 2029, total revenue would increase by 11.5%, with the pesticides market likely to reach close to US$250.96 billion. Chemicals called pesticides are used to eradicate or manage a large variety of pest species that have the potential to harm crops, animals, and agricultural productivity. Chemical agents used to manage organisms that affect crops, such as fungicides, herbicides, and insecticides, are referred to as agricultural chemicals (also known as agrochemicals). The Product Complexity Index places pesticides at position 552. Brazil, France, and the United States were the major importers of pesticides in 2021, accounting for US$36.4 billion, US$2.01 billion, and US$1.58 billion respectively. India ($2.63 billion), China ($5.25 billion), and France ($4 billion) were among the leading exporters, respectively, using pesticides in home gardens and farm areas, and so forth.

The market for nanopesticides was estimated at $0.6 billion in 2021 and projected to reach $1.9 billion by 2030, with a CAGR of 13.66% between 2022 and 2030.

Nanopesticides are employed to obtain the most advantage because they offer prolonged release kinetics with high stability, effective solubility, and increased permeability. Additionally, they aid in extending the duration of pest control efficacy and stop the breakdown of the molecules that are encased under harsh environmental circumstances.

Private equity investments in global agriculture grew from $0.8 trillion for the entire year 2012 to more over $1.8 trillion in the first half of 2013. In the fields of political economy and governance, this movement is likewise acknowledged and encouraged (Rema and Martinez, 2013). Following the recommendations of the World Development Report, in September 2009, the G20 designated agriculture expansion as a crucial activity for the advancement of developing nations (World Bank, 2103). The global economy's changing agricultural growth is being influenced by several factors led by a major group. It has been characterized as the emerging consensus among the stakeholders, development agencies, and organizations indicated in the preceding paragraphs (Ortega, 2013; World Bank, 2013).

The well-known Millennium Development goals which emphasizes on agriculture's major contribution to all eight MDGs, as well as claims made by world leaders about agriculture's future roles, clearly take into account placing agriculture's contributions and roles within the larger economic-political context of both international and regional realities having an effect on individual financial systems (De Janvry, 2009; Scoones et al., 2014).

North America, Western Europe, and China are predicted to have the most rises in food and product demand by 2050, and the commercial food industry systems will be impacted by these dynamics. Trends in population size, per capita income, aspirations for one's lifestyle, and other preferences of consumers will all have an impact on demand (Scoones et al., 2014). Consumer demand in these areas has grown and will continue to do so. It will be even more competitive, diversified, and centered on quality, product safety, and accessibility (Bairwa et al., 2014). These consumers will be interested in environmental, ethical, and social concerns in addition to seeking pleasure, surprise, and pleasant experiences in their cuisine (Hughes 2007; Vermeulen et al., 2008). According to Khodos (2014), these developments will interconnect farm production, processing, wholesaling, and retailing and are anticipated to have substantial implications on the sourcing of food items.

The choices, interests, and values of consumers emerged in these markets will have an immense effect on the agrifood value chain, as well as the associated certification requirements, traceability requirements, and monitoring systems for food safety, environmental sustainability, and ethical considerations (Raynolds et al., 2014). Changes in investment patterns are another trend that has been noticed in the agricultural industry. The restructuring of the agri-food business sector is also projected to modify the nature of agribusiness investments (Alishahi et al., 2014; Brenes et al., 2014). It has suggested that investments should concentrate on wholesale market infrastructure for combining value chain operations and enable scale economics (Raynolds et al., 2014). Therefore, rather than concentrating just on granting farm producers access to the subsequent level in the value chain, the emphasis will shift from the installation of infrastructure and support systems to facilitate the functioning of all levels within the entire agro-based chain (Rhim, Park and Ha, 2013; Khodos, 2014).

8.5 THE PROJECTED GROWTH AND FUTURE SIZE OF NANOPESTICIDE

To increase productivity per hectare, farmers all across the world are using intensive farming techniques, which increase pesticide usage. Using pesticides and chemical fertilizers, intensive farming is a technique for agricultural intensification and automation that raises yields from the available land. These approaches help to fulfil the growing population's food needs and avoid food shortages. The need for pesticides is boosted by intensive farming practices like monoculture. Monoculture refers to the practice of growing one type of crop intensively across a large area, which depletes soil nutrients and increases dependency on pesticides. In order to increase output, the demand for nanopesticides will grow in the future.

By contributing for more than 34% of yearly income generating in 2021, Europe is at the forefront in using nanopesticides. Asia Pacific was the second-largest market since it has the most arable land and has expanded acceptance of cutting-edge crop protection technologies. Additionally, it is anticipated to grow quickly during the forecast period. North America continued to have a substantial market share in 2021. The Middle East and Africa, on the other hand, are incredibly small markets, and it is probable that they will have a slower CAGR over the next years.

The research of Bukucs et al. (2014) claims that agricultural nanotechnology applications are the next industrial revolution, and both developed and developing countries are investing in this technology to attain market domination The National Nanotechnology Initiative (NNI) of the USA, which has a 4 year, 3.7 billion USD commitment, is now in the lead (Schiefer and Hortmann, 2014; Banterle et al., 2014). Japan and the European Union have both made significant financial contributions after the United States (USD 750 million and USD 1.2 billion, including individual nation payments, annually, respectively) (Sodano and Vernaeau, 2014). The level of investment in developing nations may be relatively smaller, but according to Yusuf (2014), this hasn't diminished their influence on the global scene. According to a recent report by the Helmuth Kaiser Consultancy, the market for nanofood would more than double by 2018 from 2.6 billion USD to 20.4 billion USD (Brenes et al., 2014). Over 400 companies are already involved research in the field of nanotechnology, and during the next ten years, this number is anticipated to reach over 1000. The USA is the largest economy by a wide margin, followed by China, Japan, and the EU. The market for nanotechnology was estimated by Business Communications Company, a technical market testing and industry analysis company, to be worth 7.6 billion USD in 2003 and 1 trillion USD in 2020 (BCC, 2014). The food and agricultural industries have not yet completely realized the potential of nanotechnology (Wilson et al., 2014). No area of study will be spared from the remarkable advancements brought about by nanotechnology (Dlamini et al., 2014). There is no exception in the agriculture and food industries. In order to increase the quality of agricultural products with a focus on business, nanotechnology has introduced unique methods to problems affecting livestock, plants, organic products, and post-harvest technologies (Lopez-Salazar et al., 2014). It significantly influences the creation of innovative biomaterials, the production of items, and the design of methods and instruments for assuring safety of food and biosecurity (Lopez-Salazar et al., 2014). Over the following five years, this is predicted to increase by 12.5%, reaching USD 1.9 trillion (Hirsch et al., 2014).

8.6 ECONOMIC VIABILITY

Nanopesticides are a promising alternative to traditional chemical pesticides because of their high efficacy, specificity, and potential for reduced environmental impact. However, their economic viability is still a matter of debate. On the one hand, nanopesticides are more expensive to produce than traditional pesticides due to the high cost of nanomaterials and their specialized manufacturing processes. On the other hand, nanopesticides have the potential to provide better crop protection and reduce the need for multiple applications, which can offset their higher cost (Dlamini et al., 2014). Moreover, the use of nanopesticides can lead to increased crop yields, higher quality produce, and improved food safety, which can result in higher profits for farmers and other stakeholders. Additionally, the reduced environmental impact of nanopesticides can help reduce costs associated with environmental clean-up and regulatory compliance. Overall, the economic viability of nanopesticides depends on several factors, including the cost of production, effectiveness, regulatory approval, and market demand. As the technology continues to develop, further research and commercialization efforts may help to improve the cost-effectiveness of nanopesticides and make them a more attractive option for sustainable agriculture (Hirsch et al., 2014).

8.7 CONCLUSION AND FUTURE OUTLOOK

The development of nanopesticides is a promising alternative to traditional chemical pesticides for sustainable agriculture. While there are some economic and regulatory challenges associated with their production and commercialization, several companies are actively working to bring effective and eco-friendly nanopesticides to the market. The potential benefits of nanopesticides include improved crop protection, reduced environmental impact, and increased profitability for farmers and other stakeholders. Looking to the future, the market for nanopesticides is expected to grow significantly as demand for sustainable agricultural practices increases. The increasing awareness of the negative impact of traditional chemical pesticides on human health and the environment is driving the adoption of alternative pest control methods, including nanotechnology-based solutions. Furthermore, advances in nanotechnology and materials science will likely lead to the development of new and more effective nanopesticides. The economical and market status of nanopesticides is promising, but further research and investment are necessary to fully realize their potential. Regulatory approval, cost-effectiveness, and market demand will continue to be important factors in determining the success of nanopesticides in the future.

KEYWORDS

- **agrifood**
- **nanofood**
- **nanopesticides**
- **nanosulfur**
- **nanotechnology**
- **toxicity**

REFERENCES

ACS. (2014). Special symposium on the "fate, effects and risks of nanopesticides." In *National Meeting of the American Chemical Society* (San Francisco, CA).

Alishahi, A., Proulx, J., & Aider, M. (2014). Chitosan as biobased nanocomposite in the seafood industry and aquaculture. In *Seafood Science: Advances in Chemistry, Technology and Applications* (p. 211). https://doi.org/10.1201/b17402-12.

Aschberger, K., Gottardo, S., Amenta, V., Arena, M., Moniz, F. B., Bouwmeester, H., et al. (2015). Nanomaterials in food–Current and future applications and regulatory aspects. *Journal of Physics: Conference Series*, *617*(1), 1–6. https://doi.org/10.1088/1742-6596/617/1/012032.

Australian Pesticides and Veterinary Medicines Authority (APVMA). (2014). A draft report from the Australian pesticides and veterinary medicines authority: Regulatory considerations for nanopesticides and veterinary nanomedicines. http://apvma.gov.au/sites/default/files/docs/report-draft-regulatory-considerations-nanopesticides-veterinary-nanomedicines.pdf.

Bairwa, S. L., Kalia, A., Meena, L. K., Lakra, K., & Kushwaha, S. (2014). Agribusiness management education: A review on employment opportunities. *International Journal of Scientific and Research Publications*, *4*(2), 1–4.

Banterle, A., Cavaliere, A., Carraresi, L., & Stranieri, S. (2014). Food SMEs face increasing competition in the EU market: Marketing management capability is a tool for becoming a price maker. *Agribusiness*, *30*(2), 113–131. https://doi.org/10.1002/agr.21354.

Brenes, E. R., Montoya, D., & Ciravegna, L. (2014). Differentiation strategies in emerging markets: The case of Latin American agribusinesses. *Journal of Business Research*, *67*(5), 847–855. https://doi.org/10.1016/j.jbusres.2013.07.003.

Crop and Chemical Europe. (2015). Agrochemical formulation. In *Conference organized by Informal Life*. http://www.informa-ls.com/event/cropform2014/Content (accessed on 3 October 2024).

De Azeredo, M. C. H. (2013). Antimicrobial nanostructures in food packaging. *Trends in Food Science & Technology*, *30*, 56–69. https://doi.org/10.1016/j.tifs.2012.11.006.

Dlamini, B. P., Kirsten, J. F., & Masuku, M. B. (2014). Factors affecting the competitiveness of the agribusiness sector in Swaziland. *Journal of Agricultural Studies*, *2*(1), 61–72. https://doi.org/10.5296/jas.v2i1.4775.

ETC Group. (2004). Down to the farm: The impact of NANO-scale technologies on food and agriculture. http://www.etcgroup.org/ (accessed on 3 October 2024).

European Commission (EC). (2014). JRC Nanomaterials repository. https://ec.europa.eu/jrc/en/scientific-tool/jrc-nanomaterials-repository (accessed on 3 October 2024).

European Commission. (2009). Regulation EC 1107/2009 concerning the placing of plant protection products on the market: Guidance on section 5–Fate and behavior in the environment. http://eur-lex.europa.eu/legal-content/EN/TXT/?uri=celex:32009R1107 (accessed on 3 October 2024).

European Food Safety Authority (EFSA). (2009). Updating the opinion related to the revision of Annexes II and III to Council Directive 91/414/EEC concerning the placing of plant protection products on the market–Toxicological and metabolism studies. *EFSA Journal*, *1166*, 1–6. https://doi.org/10.2903/j.efsa.2009.1166.

European Union (EU). (2011). Commission recommendation of 18 October 2011 on the definition of nanomaterial. http://ec.europa.eu/environment/chemicals/nanotech/#definition (accessed on 3 October 2024).

Food and Agriculture Organization of the United Nations & World Health Organization (FAO/WHO). (2010). *Expert Meeting on the Application of Nanotechnologies in the Food and Agriculture Sectors: Potential Food Safety Implications*. http://whqlibdoc.who.int/publications/2010/9789241563932_eng.pdf.

Food and Agriculture Organization of the United Nations & World Health Organization (FAO/WHO). (2013). *State of the Art on the Initiatives and Activities Relevant to Risk Assessment and Risk Management of Nanotechnologies in the Food and Agriculture Sectors*. http://www.fao.org/docrep/018/i3281e/i3281e.pdf (accessed on 3 October 2024).

Friends of the Earth. (2008). Out of the laboratory and on to our plates: Nanotechnology in food and agriculture. http://www.foe.org/system/storage/877/b5/4/547/Nanotechnology_in_food_and_agriculture_-_web_resolution.pdf (accessed on 3 October 2024).
Gewin, W. (2015). Everything you need to know about nanopesticides. *Modern Farmer*. http://modernfarmer.com/2015/01/everything-need-know-nanopesticides/ (accessed on 3 October 2024).
Gogos, A., Knauer, K., & Bucheli, T. D. (2012). Nanomaterials in plant protection and fertilization: Current state, foreseen applications, and research priorities. *Journal of Agricultural and Food Chemistry*, *60*(39), 9781–9792. https://doi.org/10.1021/jf302154y.
Hirsch, S., Schiefer, J., Gschwandtner, A., & Hartmann, M. (2014). The determinants of firm profitability differences in EU food processing. *Journal of Agricultural Economics*. https://doi.org/10.1111/1477-9552.12061.
Joint Research Centre–Institute for Prospective Technological Studies (JRC-IPTS). (2014). *Nanotechnology For the Agricultural Sector: From Research to The Field*. https://ec.europa.eu/jrc/sites/default/files/ipts_jrc_89736_(online)__final.pdf (accessed on 3 October 2024).
Kah, M., & Hofmann, T. (2014). Nanopesticides research: Current trends and future priorities. *Environment International*, *63*, 224–235. https://doi.org/10.1016/j.envint.2013.11.015.
Kah, M., Beulke, S., Tiede, K., & Hofmann, T. (2013). Nanopesticides: State of knowledge, environmental fate and exposure modelling. *Critical Reviews in Environmental Science and Technology*, *43*(16), 1823–1867. https://doi.org/10.1080/10643389.2012.671750.
Kah, M., Machinski, P., Koerner, P., Tiede, K., Grillo, R., Fraceto, L. F., et al. (2014). Analyzing the fate of nanopesticides in soil and the applicability of regulatory protocols using a polymer-based nanoformulation of atrazine. *Environmental Science and Pollution Research*, *21*(18), 11699–11707. https://doi.org/10.1007/s11356-014-2523-6.
Khodos, D. (2014). Economic mechanism of innovative development of the agricultural sector. *Innovation Management and Company Sustainability*, *26*, 73.
Kookana, R. S., Boxall, A. B. A., Reeves, P. T., Ashauer, R., Beulke, S., Chaudhry, Q., et al. (2014). Nanopesticides: Guiding principles for regulatory evaluation of environmental risks. *Journal of Agricultural and Food Chemistry*, *62*(19), 4227–4240. https://doi.org/10.1021/jf500232f.
Lauterwasser, C. (2005). Small sizes that matter: Opportunities and risks of nanotechnologies. Report by Alliance in co-operation with the OECD International Futures Programmes. http://www.oecd.org/science/nanosafety/37770473.pdf (accessed on 3 October 2024).
Liu, R., & Lal, R. (2015). Potentials of engineered nanoparticles as fertilizers for increasing agronomic productions. *Science of the Total Environment*, *514*, 131–139. https://doi.org/10.1016/j.scitotenv.2015.01.104.
López-Salazar, A., López-Mateo, C., & Molina-Sánchez, R. (2014). What determines the technological capabilities of the agribusiness sector in Mexico? *International Business Research*, *7*(10), 47. https://doi.org/10.5539/ibr.v7n10p47.
Rama, R., & Martínez, C. (2013). The changing structure of the global agribusiness sector. *Handbook of Industry Studies and Economic Geography*, *305*. https://doi.org/10.4337/9781782549000.

Raynolds, L. T., Long, M. A., & Murray, D. L. (2014). Regulating corporate responsibility in the American market: A comparative analysis of voluntary certifications. *Competition & Change*, *18*(2), 91–110. https://doi.org/10.1179/1024529414Z.00000000050.

Scoones, I., Smalley, R., Hall, R., & Tsikata, D. (2014). Narratives of scarcity: Understanding the "global resource grab."

Scott, N., & Chen, H. (2002). Nanoscale science and engineering for agriculture and food systems. Report submitted to Cooperative State Research, Education and Extension Service of the US Department of Agriculture. http://www.nseafs.cornell.edu/web.roadmap.pdf (accessed on 3 October 2024).

Sekhon, B. S. (2014). Nanotechnology in agri-food production: An overview. *Nanotechnology, Science and Applications*, *7*, 31–53. https://doi.org/10.2147/NSA.S39406.

Society of Environmental Toxicology and Chemistry (SETAC). (2014). Keynote on "nanopesticides and veterinary nanomedicines: Regulatory challenge?" In *SETAC Asia/Pacific Conference* (Adelaide, AU).

Suppan, S. (2013). Nanomaterials in soil: Our future food chain? Institute for Agriculture and Trade Policy. http://www.iatp.org/documents/nanomaterials-in-soil-our-future-food-chain

U.S. Department of Agriculture (USDA). (2015). U.S. Department of Agriculture awards $3.8 million in grants for nanotechnology research. http://nifa.usda.gov/press-release/usda-awards-38-million-grants-nanotechnology-research (accessed on 3 October 2024).

Vive Crop Protection. (2012). http://www.vivecrop.com/ACCN_April2012.pdf (accessed on 3 October 2024).

Watson, S. B., Gergely, A., & Janus, E. R. (2011). Where is "agronanotechnology" heading in the United States and European Union? *Natural Resources & Environment*, *26*, 8–12.

Wilson, W. W., Shakya, S., & Dahl, B. (2014). Dynamic changes in spatial competition for fertilizer. https://doi.org/10.1016/j.agsy.2014.11.006.

World Bank. (2013). *Growing Africa: Unlocking the Potential of Agribusiness*. AFTFP/AFTAI. World Bank: Washington, DC.

Yusuf, S. A. (2014). The analysis of export performance of newly industrialized countries (NICs): The lesson for African countries.

CHAPTER 9

Environmental Risk Assessment of Nanopesticides

PAVAN KUMAR GANECHARY,[1] TANISHA GUPTA,[1] and CHANDRA SHEKHAR YADAV[2]

[1]*Laboratory of Biology and Biotechnology, National Forensic Sciences University, Gandhinagar, Gujarat, India*

[2]*Forensic Biology and DNA Forensic Lab. National Forensic sciences University, Guwahati Campus, Assam, India*

ABSTRACT

The development of nanopesticides is one of the best technological innovations for the agricultural sector. As nanopesticides possesses high water solubility, safeguards agricultural compounds against environmental deterioration, and prevents weeds, insects, and infections enhancing crop production. A number of nano-formulations such as nanocapsules (e.g., with polymers) and emulsions (e.g., nanoemulsions), and there are organic as well as inorganic nanopesticides such as active ingredients, polymers and oxides of metals in micelles and particles are available in the market. The rising interest in the utilization of nanopesticides brings up questions regarding how to examine the environmental risk of the nanopesticides for usual reasons. Unexpected risk of nanopesticides has been reported for their possible toxicity to organisms which are non-targeted and from getting interacted with other compounds present in the environment. It has been found that nanopesticides may result in toxicological effects of various flora and fauna because of their property of biomimetics, high ability of distribution (dispensation) and bioaccumulation in local water

Nanopesticides: A Perspective from Welfare to Risk Assessment.
Garima Awasthi, Payal Lodha, and Gajendra Pal Singh (Eds.)

DOI: 10.1201/9781003605874-9

bodies, soil, food materials and which consequently affect the lifecycle of organisms such as animals mainly in soil bacteria, non-targeted organisms such as Earthworms, plants and mammals like humans, that includes reduced cell number, structural damage, reduced glomalin concentration. In this chapter we have discussed the nanopesticides induced toxicity in general as well as ongoing methodologies for finding risk of nanopesticide exposure in the environment (Figure 9.1).

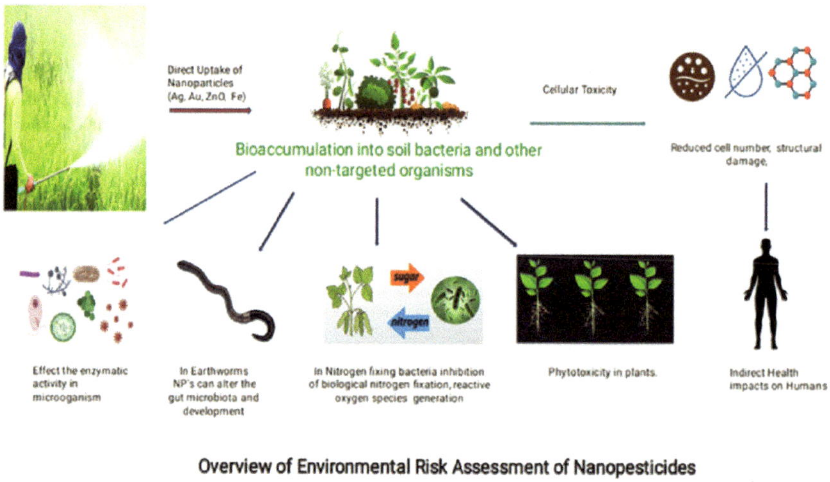

FIGURE 9.1 Overview of environmental risk assessment of nanopesticides.

9.1 INTRODUCTION TO NANOPESTICIDES

Main drawbacks of currently used insecticides, pesticides and herbicides are that they are frequently composed of extremely insoluble compounds. For adequate potency and consistency solubility of these compounds is must for application in the fields, but in order to increase their solubility addition of organic solvents are required. Addition of solvents increases the cost as well as create a constant risk to the applicator and other secondary targets. Unfortunately, an excessive reliance on these inventions has had several negative impacts. Eutrophication, nitrate pollution, and micronutrient imbalance are all harmful effects of using fertilizers too much or incorrectly. Similarly, non-targeted animals like humans, helpful microbes, beneficial insects, birds, earthworms, and aquatic life forms are most at risk from extensive pesticide use (Ameen et al., 2021). The major issues connected to the usage of pesticides include biomagnification, the emergence of pesticide resistance, and the return of pest populations. Chemical pesticides need to have their active components spread out in a liquid phase in many applications in order to make spraying easier.

However, the fact that in water, pesticides are either barely soluble or fully insoluble are their principal drawbacks. This is typically due to high lipophilicity (nonpolarity), it is thermodynamically unfavorable to mix it with water (The interfacial free energy increases as the contact surface area between these incompatible phases grows). In particular cases, pesticide has a high lattice energy and strong intermolecular bonds in the solid form (indicated by a melting point greater than 200°C).

As a result, unless constant strong agitation is carried out, the processes of flocculation, coalescence, sedimentation, Ostwald ripening and creaming cause these pesticide droplets or particles to separate from water as shown in Figure 9.2. This separation makes it considerably more challenging to evenly apply the pesticide to treated surfaces, necessitating the use of significantly larger amounts of the pesticide to guarantee that an adequate dosage is present in every region to control the target species or organism. This increases costs and worsens the impact of such pesticides on the surroundings as well as on human health. Pesticide components can

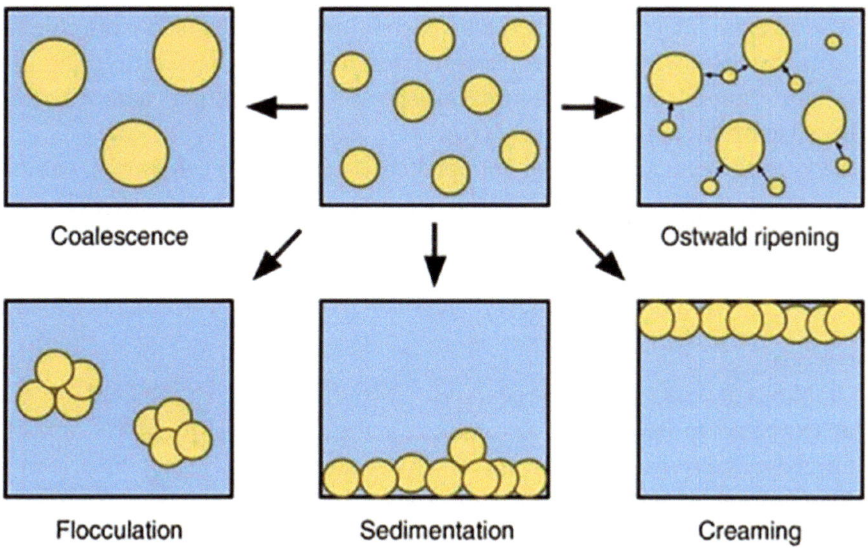

FIGURE 9.2 Illustration of the demixing process for non-polar pesticide droplets that have been distributed.

have their particle sizes reduced to nano or colloidal scales using nanotechnology to dramatically increase their water solubility, rate of dissolution, and homogenous dispersion when applied. Everyone's health is at risk if we practice the indiscriminate and improper application of pesticides to impact the equilibrium of ecosphere. Adverse effects of accidental ingestion of pesticide residue for long term or short term from drinking water, food results in disabilities and lethality. The category of generation which is more exposed to this is children which may result in their permanent organ and tissue damage.

9.1.1 WHAT ACTUALLY ARE NANOPESTICIDES

Agrochemical components are transported via nanopesticides, which are two- to three-dimensional nanostructures with diameters between 1 and 200 nm (active ingredients). Because of their distinguishing characteristic, the incorporation of active ingredients into nanoparticles provides advantages over free insecticides. Pesticides that have been particularly anchored on a hybrid substrate, enclosed in a matrix and/or functionalized as nanocarriers for triggering by environmental factors or enzyme-mediated factors are referred to as "nanopesticides." There is optimism that nanopesticides will offer novel, cutting-edge nano-based formulations that are durable and agile in various spray environments (sun, heat, rain), penetrate organisms (virus, fungus, insects), can be transported to the target, and withstand defense of the pests or disease (Kah et al., 2016).

Predominantly speaking, nanopesticides are of two categories, namely:

1. **Inorganic Active Ingredients (AI):** It is at the nanoscale, such as $Cu(OH)_2$ nanopesticides, without carriers;
2. **Active Ingredient (AI):** Nanocarrier organic molecular complexes or polymers.

Nevertheless, any nanopesticide formulation is anticipated as well as expected to be able to increase AI stability, extend the duration of its efficacy, and reduce the pesticides' environmental burdens. China is more than any other country likely to overtake the United States of America as the world's top producer and user of nanopesticides as spending on agricultural nanotechnology research and development rapidly rises. Regulation and scientific assessment of nanopesticides are therefore urgently needed.

Innovative formulations of nanocarriers based on various materials including metal, carbon, silica, lipids, polymers, copolymers and ceramic are anticipated to investigate pesticide actions by the combination of their structure and unique features. By enhancing some properties such as water solubility, bioavailability, and shielding agrochemicals from environmental deterioration, the adoption of nanopesticide formulations will revolutionize crop disease, weed, and insect management. A huge number of different types of formulations have been available such as nanocapsules (e.g., nanopolymers) emulsions (e.g., nanoemulsions), and there are organic as well as inorganic nanopesticides such as AI polymers and oxides of metals in micelles and particles.

Using polymeric materials that are very sensitive to light, temperature-sensitive, humidity-sensitive, specific enzyme-sensitive, and soil pH-sensitive, nanocapsules for insecticides are available in the market. There are numerous ways to create these nanostructured materials, including adsorption, coupling, encapsulation, and embedding.

Because nanoparticles operate differently from dissolved chemicals and have distinct environmental dangers than traditional pesticides, nanopesticides are one of the finest technology advancements to be introduced to the market with formulation improvements.

9.2 APPLICATIONS OF NANOPESTICIDES

The thriving population is demanding a surge in the production of crops all over the world but, this increases in the pest problem and the plant pathogens. This further creates pressure on both the Government and Non-Government agencies to adapt new technologies for achievement of the production goals. As said by the United Nations, on an average of 2,00,000 persons die per year from toxic exposure to pesticides.

Nanotechnology is now demonstrating a broad range of growth in several industries, including medical, the petroleum sector, electronics, chemistry, and agriculture. A nanoparticle is an extremely small particle that ranges in size from one to one hundred nanometers. Although undetectable to the naked eye, nanoparticles exhibit extremely different physical and chemical properties from their bulk materials. A material's properties change as it gets closer to the atomic scale.

Microencapsulated pesticides better known as nanopesticides include fungicides, insecticides, herbicides, and others. By using more modern and sophisticated formulations that can enter insect bodies, nanopesticides leverage nanotechnology to protect plants and fight insect pests. The formulations of nanopesticides can alter the management of weeds, diseases, and insects in crops by increasing agrochemicals' bioavailability, water solubility, and resistance to environmental degradation.

The nanopesticides market was evaluated to be worth $0.5 billion in the year 2021, and it is anticipated to grow to $1.6 billion by the year 2031, with a CAGR of 12.5% between 2022 and 2031. Harvesting, production, protection, and packaging are the four main segments that make up the nanopesticides market. This is mainly due to their greater pest control capabilities compared to conventional pesticides, nanopesticides are much sought after in agriculture. Encapsulated pesticides are also becoming more and more common in a variety of end-user industries, including food processing, storage, and packaging. By contributing more than 34% of yearly income generating in 2021, Europe is at the forefront of the use of nanopesticides. Asia Pacific was the second-largest market since it has the most fertile land suitable for agriculture and has expanded acceptance of cutting-edge crop protection technology (Nanopesticides Market Size, Growth, Trends, Report 2022–2030, n.d.). Additionally, it is anticipated to grow quickly during the forecast period.

Type, utility, application are the segments used to categories the nanopesticide market and is further divided into nano-insecticides, nano-herbicides, nano-fungicides, and others based on kind. The market is divided into food crops and industrial crops based on utility as well as into pest control, packaging and other categories based on application.

Nanopesticides can be used to resolve abiotic-type stress created by traditional pesticides since they can provide protection against them. The active chemical ingredients of pesticides are sustained by encapsulation, which also protects them against environment degradation and lowers the cost of pesticides by reducing waste. The formulation of nanopesticides enhances the properties and behaviour of pesticides, like the stability, motility, dispersion, dissolving, and targeted delivery. Additionally, it can improve the efficiency, affordability, and safety of traditional pesticides. This is accomplished by extending the time an effect lasts, increasing effectiveness, reducing the dose, managing the release of the active components, and enhancing the stability of payloads from the environment, which in turn reduces environmental residuals. Improving the precision and accuracy of active ingredient administration will result in improvements in efficacy, longevity, and a reduction in the amount of active compounds required.

Additionally, by reducing the amount of material released into the ecosystem, this will lessen any unwanted effects on species that are not the target. A variety of formulation kinds, such as nanoemulsions, nanocapsules (containing polymers), and inorganic Engineered Nanoparticles (ENPs), such as metal oxides, metals, and nanoclays, have been proposed (Pradhan and Mailapalli, 2017). Nanopesticides need to offer a number of benefits in order to be utilised to successfully protect crops from insects and illnesses (including greater productiveness and durability, great dispersion and hydrophilicity, capacity to biodegrade in the environment and soil, and lack of toxic effects, and photo generative nature). Researchers are working to create nanopesticides that are more effective while also being less hazardous to the environment due to the rapid advancements in nanopesticide research. Therefore, target-specific nanopesticides ought to aid in minimising the harm to non-target crops and the quantity that goes into the environment.

Nanopesticides are utilised in industrial crops, including cotton, tobacco, and jatropha. In industrial crops, nanopesticides are employed as fungicides to eradicate soil-borne diseases, herbicides to eradicate weeds and soil fumigants, and insecticides to eliminate plant-eating pests. Industrial crop management techniques such as sowing, plant density, fertilizer, pruning, shading, management of wild stands for sustainable harvest, insect and weed control, harvest, post-harvest, and others boost production and particular chemical components. In wasteland phytostabilization, industrial crop cultivation is anticipated to reduce erosion, leaching, and groundwater contamination while enhancing soil fertility and structure.

9.3 NANOPARTICLES INDUCED TOXICITY

The safety of nanoparticles is a growing concern since even products that benefit biodiversity may have undetected adverse consequences on the environment at very low concentrations. If nanoparticles (NPs) interact with biological molecules or chemical components of living organisms or build up in the environment, direct and indirect health risks may result. Environmental assessment of nanopesticides is very difficult because of the dynamic behaviour of nanoparticles in the natural environment and the resulting changes in their physicochemical properties. For those slow/targeted release nanoformulations in particular, these features make it challenging to collect enough data to enable exposure modelling.

While the variety of microbial species in the rhizosphere (the layer of soil surrounding a plant root where the biology and chemistry of the soil are largely influenced by the root) of ectomycorrhizal trees might expand, the community's makeup between specialized and generalist root symbiotic bacteria might vary. In nutrient- and water-limited environments, these changes may have an unfavorable impact on the health of the trees. This might also apply to microbes that encourage plant development. The distribution of herbivorous and predatory animals may change over time, which could have unexpected ecological effects.

Similar to this, some microbial species can thrive and flourish when exposed to low concentrations of nanomaterials, however large quantities can cause cell death. Because the effects of nanopesticides are species-specific, low concentrations/doses changes the species composition and/or richness in the ecosystem without regard for non-target consequences.

The understanding of the toxicity, specificity and genotoxicity of nanoparticles in humans has been vastly studied that deals with the harmful toxic effects and dangers connected with nanoparticles, anionic nanoparticles, solvents, surfactants and inducing agents. However, when active ingredients and nanotechnology are combined, the toxicological risk of achieving all objectives appears to increase. When delivering active ingredients with limited water solubility, lipid-based formulations including liposomes, micelles and liquid crystals are typically formed of natural cholesterol, phospholipids and glycolipids. This requires the inclusion of surfactants and may enhance the toxicity of insecticides.

Children are more prone to permanent tissue damage and organ system damage from pesticide exposure. The combination of the central and peripheral neurotoxicity as well as the impact on the blood's capacity to clot are significant causes for worry. In fact, for the safe, prolonged and sustainable expansion of use of nanoparticles (nanopesticides) in agricultural industry, a thorough analysis of the benefits and drawbacks that affect the activity and toxicity of nanopesticides is very essential as shown in Figure 9.3. The impact of formulations on how nanopesticides behave in habitats, ecosystems, agricultural labour, consumers, and every other productive zone or region engaged in the agriculture chain is not fully acknowledged and understood. However, it is well recognized that nanoformulations play a pivotal role in minimising the degradation of active chemicals, enhancing water solubility equilibrium, and enhancing the biological availability of active substances. To be more precise, nanoformulations prevent endemic pest infestation, total plant damage, and financial losses by lowering the quantity of agricultural goods and foods that are squandered or splurged.

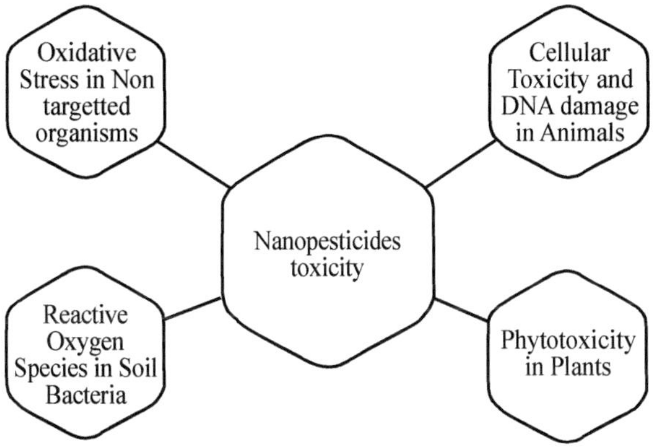

FIGURE 9.3 General overview of toxicity produced by nanopesticides.

Nanoscale active ingredients (AI) or carriers will enter the natural environment after being exposed to nanopesticides, where they are likely to change over their life cycles. Additional or unexpected effects of nanoscale components in the nanopesticides have been observed in addition to the traditional exposure situations, necessitating further research. For instance, when compared to traditional $CuSO_4$, $Cu(OH)_2$ nanopesticides, whether they were pure (AI) active ingredient or commercial formulations, significantly slowed down the degradation of thiacloprid – extremely problematic insecticide, extending its residual persistence in soil environment and potentially increasing the risk to non-target organisms (Zhang et al., 2019).

The metabolic reactions of living organisms are the main focus of another significant environmental danger associated with nanopesticides. For instance, the metabolite profiles of spinach plants treated with conventional $CuSO_4$ or $Cu(OH)_2$ nanopesticide both significantly changed; in particular, the nitrogen metabolism disruption was substantially nanopesticide-specific (Zhao et al., 2017). For organisms, metabolic abnormalities pose serious physiological problems.

Studies of the various biochemical reactions of nanopesticides compared to their conventional analogues are very much essential in order to usher future research addressing the environmental risks associated with nanopesticides (Pandey et al., 2016). In order to produce a more accurate evaluation of the behaviour and risk of nanopesticides and their conventional counterparts in the natural environment, additional research, including long-term sampling in the subsequent years after application, is required to bridge the well-known openings or clefts between laboratory experiments that are highly simplified and real-world agricultural field settings (Kah et al., 2018).

AgNPs have recently been found to be size-dependently hazardous to human cell lines, which may be linked or correlated to oxidative stress. As was found out in a study correlating to the testing of Silver nanoparticles (AgNPs) of size 47 nm and 42 nm as well as Gold nanoparticles (AuNPs) of size 30 nm, 50 nm, and 90 nm in human pulmonary fibroblasts (HPF) found that 47 nm AgNPs were more harmful than the bigger AgNPs and AuNPs, but equivalent cytotoxicity was seen for entire set of AuNPs. A large rise in ROS generation, a drop in glutathione levels, and a minor inactivation of superoxide dismutase were all indicators of the development of oxidative stress.

AgNPs are produced at a rate of 320 tonnes annually and are widely employed in numerous sectors such as textile coatings, health industry, food storage, and a horizon of applications of the environment due primarily to their antibacterial qualities.

The water supply becomes contaminated with nanopesticides from agricultural and industrial wastewater runoff after a precipitation event due to a phenomena known as soil penetrating leaching, which raises issues for the environment. The respiratory, cardiovascular, lymphatic, autoimmune, neurologic, and different cancers might present acutely after exposure or years later due to bioaccumulation and unique nanoparticle features (Deng et al., 2017). These adverse effects are connected to individual vulnerability (susceptibility) and exposure time to nanoparticles. The following sections cover a few of the most important topics.

9.3.1 NANOPESTICIDES INDUCED TOXICITY IN SOIL BACTERIA

Application of nanopesticides has shown an increase in crop production. However, concern has been raised over the effects of nanomaterials on health of the plants, soil organisms and microorganisms with reference to even food quality and soil fertility. However, the majority of research is presently conducted in laboratory settings that are oversimplified (such as hydroponics or soil microcosms), at high concentrations, or with pristine nanoparticles that are incomparable to market nano-enabled goods. While this nanopesticide may shield crops against bacterial and fungal illnesses, it may also adversely affect non-target plant-associated microorganisms that are essential to plant nutrition, such as mycorrhizal communities of bacteria that fix atmospheric nitrogen as described in Figure 9.4. Moreover, broad-spectrum nanopesticides may be toxic to helpful soil microbes that break down organic matter (OM), which prolongs and sustains long-term fertility of the soil layer.

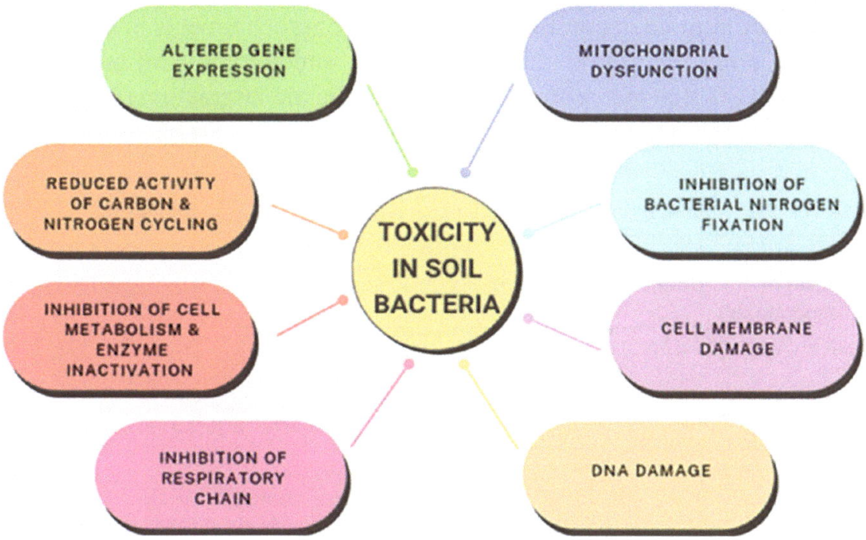

FIGURE 9.4 Depiction of toxicity generated by nanoparticles in soil bacteria.

The immobilization, cycling, and degradation of pollutants, along with the detoxification and transformation of nutrients and carbon by soil microorganisms, ultimately contribute to improved soil health. Approximately 15% of the total microbial communities in widely distributed heterotrophic microflora consist of bacterial populations from various species that promote plant growth directly or indirectly. These rhizosphere-dwelling bacterial communities, known as plant growth-promoting rhizobacteria (PGPR), have the ability to colonize plant roots. Prominent PGPR species that promote plant growth include *Acinetobacter*, *Agrobacterium*, *Arthrobacter*, *Azospirillum*, *Azotobacter*, *Bacillus*, *Bradyrhizobium*, *Burkholderia*, *Pseudomonas*, *Rhizobium*, *Serratia*, and *Thiobacillus*.

Some nanoinsecticides, such as pyraclostrobin, have shown potential to boost rice crop yields affected by rice blast fungus, though they have been found to be hazardous to aquatic species. The photolytic degradation rate of pyraclostrobin in water can be reduced by encapsulating the active ingredient. For instance, Seltima, a microcapsule formulation, minimizes pyraclostrobin's toxicity by releasing it gradually and directly onto the rice plant's leaves. Similarly, embedding abamectin in hollow SiO_2 nanospheres can improve its stability and residual activity, providing controlled release.

Regular exposure to these nanopesticide treatments may have unforeseen effects on soil fertility and plant yields over time. Typically, nanopesticides are applied in combination with other agrochemicals in agricultural soils or crops, and these combinations can interact with soil microorganisms in ways that may either reduce (antagonistic interactions) or enhance (additive interactions) the overall effects, as shown in Table 9.1.

TABLE 9.1 Nanoparticles and Their Effects on Various Soil Microorganisms

Nanoparticles Present in Nanopesticides	Size (nm)	Soil Microorganism	Effect	References
Silver (Ag)	10–50	*Azotobacter vinelandii*	Reduction in cell number, generation of reactive oxygen species (ROS), induced apoptosis, inhibition of biological nitrogen fixation (BNF), structural damage	Zhang et al. (2018)
Zerovalent Silver (Ag)	10	*Nitrosomonas europaea*	Cell wall damage, disintegrated nuclei, and size-dependent reduction of NH_3 oxidation	Arnaout and Gunsch (2012)
Zerovalent Iron (Fe)	20–90	*Paracoccus* sp.	Enhanced cell growth, NO_3^- biodegradation, and reduction in cell density	Liu et al. (2014)
Copper Oxide (CuO)	35	Soil nitrifying bacteria	Suppression of nitrification kinetics due to CuO nanoparticles oxidizing to hydroxides and Cu(II)-oxide under toxic conditions with increasing aging time	VandeVoort and Arai (2018)
Zinc Oxide (ZnO)	10–30	Aquatic microcosm	Negative impact on fungal community structure and microbial enzyme activities	Wu et al. (2020)
Silver (Ag)	0–10	*Vicia faba* and *Rhizobium leguminosarum*	Slowed symbiosis and soil microbial activity; significant reduction in glomalin concentration, nodulation, mycorrhizal colonization, and nitrogenase activity	Abdelkhalek et al. (2022)

The symbiotic relationship between soil-grown *Vicia faba* and *Rhizobium leguminosarum*, *Glomus aggregatum*, or a combination of both cultures has been shown to be affected by silver nanoparticles (AgNPs), a crucial component in some nanopesticides. At a concentration of 800 mg kg^{-1}, AgNPs reduced both symbiosis and soil microbial activity. AgNPs also significantly lowered glomalin concentration, nodulation, mycorrhizal colonization, and nitrogenase activity. Furthermore, exposure to Ag-NPs led to the breakdown of bacteroids and the intracellular degradation of cytoplasmic components through autophagy.

In a separate study, it was found that 306 mM ZnO nanoparticles (ZnONPs) completely inhibited the growth of *Pseudomonas* species, while 7.7 mM ZnONPs of a size 100 nm reduced colony-forming units of *Pseudomonas chlororaphis* 06 by 12.5% (Ameen et al., 2021). The garden pea and its symbiotic bacterium, *Rhizobium leguminosarum*, were used to assess the impact of ZnONPs on the rhizobium-legume symbiosis. Exposure to ZnONPs led to morphological alterations in rhizobial cells and damage to the bacterial surface. ZnONPs also accelerated nodule senescence, delayed the initiation of nitrogen fixation, and disrupted the root nodulation process. The study concluded that the environmental presence of ZnONPs and their interaction with beneficial soil bacteria could adversely affect crop yield (Ameen et al., 2021).

9.3.2 NANOPESTICIDES INDUCED TOXICITY IN PLANTS

Upon entering a plant's cell wall following treatment, nanopesticide particles are absorbed. The plant cell wall functions as a semipermeable environmental barrier, regulating the passage of external substances through its pores and across the cell membrane. Therefore, the size, charge, and physicochemical properties of nanopesticides are key factors in their ability to pass through these pores. Nanopesticides smaller than the diameter of the plant cell wall's pores can easily penetrate the cell wall and plasma membrane. Accordingly, nanopesticides are likely to enter plant cells via endocytosis, ion channels, carrier protein binding, or by forming complexes with transporters on the plasma membrane, as illustrated in Figure 9.5.

Once inside a cell's cytoplasm, nanopesticides can alter the plant's biochemical processes. Plants can absorb active nanopesticide components (especially metals), which may then bioaccumulate throughout the food chain. Enzymes can expedite the breakdown of nanoparticles, organic compounds, detrital inputs, pesticides, and nutrients at the molecular level, enabling their recovery for use by plants and other organisms (Zulfiqar et al., 2019). Clearly, nanotechnology has the potential to enhance agricultural sustainability; however, the cost-benefit analysis of Nanoparticle Fertilizers (NPFs) should consider soil bacteria and enzymes as critical factors in assessing ecological risk.

Inside the cells nanopesticides may interact as well as associate with proteins, DNA, organelles of the cell, hormones, and other substances and thus influence the physiological or metabolic functions at a specific location. A charged nanoparticle may cause the lipid membrane to develop pores, which would weaken the stability of the membrane and increase toxicity. The mechanical impacts of the nanopesticides on plants are connected to their size and form, which confines them to a certain location on the surface of the plant or organ. However, the impacts are sensed throughout the vegetal body and either promote or prevent plant growth. Carbon nanotubes (CNTs) having size 3 nm, for example, can enhance tomato's germination owing to its impact on water content and its entry.

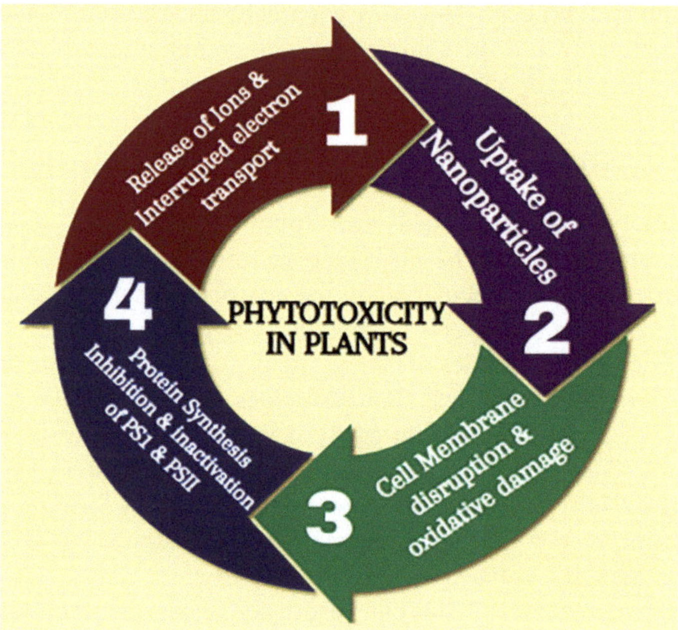

FIGURE 9.5 Mechanism of phytotoxicity in plants.

Contrarily, carbon-based nanoparticles might produce an inhibition of growth due to mechanical damage brought on by a perforating action on membranes in cells or organelles. The physicochemical characteristics of the nanomaterial determine how nanopesticides affect plants.

Plants may react differently to characteristics like different type, concentrations, free charges on the nanoparticle's surface, or created complexes; these impacts may be grouped into the following categories:

- **Favourable Response:** Inducing cellular elongation, tolerance, and the development of resistance over biological and/or abiotic stress in the environment as a favourable response to the germination process,
- **Inhibitory Response:** As a growth inhibition that is unfavorable and linked to the direct phytotoxic effects of nanoparticles or from nanopesticide, leading to the excess production of reactive oxygen species (ROS), altering normal functional cell equilibrium and its metabolic stability.

The term ROS refers to a group of naturally occurring cellular oxidative metabolic byproducts that include free radicals like ions (OH and OH^-), superoxide (O_2), hydroxyl radicals nonradicals like hydrogen peroxide (H_2O_2). To keep cells in a state of equilibrium, ROS levels within the cells must be low. Otherwise, excess ROS can be extremely harmful for plants at high doses and can cause DNA damage, lipid peroxidation, enzyme inhibition, and cell death.

Although high ROS levels have a number of detrimental impacts on plant cell metabolism, these molecules may also be involved in intricate regulatory systems that integrate the plant and its environment. The balance between ROS formation and elimination of their excess by cellular antioxidant molecules determines the beneficial role of ROS in plant control. For instance, metallic nanoparticles have a special ability to stimulate the production of ROS inside of cells. This is because most oxidants and free radicals on surface of nanoparticles can undergo Fenton reactions to form ROS.

Controlling the nanomaterials' physicochemical characteristics, like their chemical properties, concentration, size (1–100 nm), shape (irregular or geometrically defined), grouping (aggregation), including the sensitivity of the target plant species and other surface chemical properties, can enhance the targeting, cellular uptake, and intracellular trafficking of nanopesticides as shown in Table 9.2. Depending on the properties of nanomaterial or pesticide's method of action, many pathways are involved in the mechanisms of oxidative stress and plant growth suppression caused by nanopesticides as a whole.

TABLE 9.2 The Effect of Size of Nanoparticles on Various Species of Plants

Nanoparticle Size (nm)	Species	Effect on Plant	References
AgNPs – 10 nm	Thale cress (*Arabidopsis thaliana*)	Inhibition of root growth; decline in chlorophyll a, chlorophyll b, and total chlorophyll	Rastogi et al. (2017)
AgNPs – 10 nm	Common wheat (*Triticum aestivum*)	Reduction in shoot and root length; buildup of oxidized glutathione, indicating ROS formation	Dimkpa et al. (2013)
AgNPs – 20 nm	Corn (*Zea mays*)	Improved quantitative yields of fodder maize	Berahmand et al. (2012)
AgNPs – 25 nm	Rice (*Oryza sativa*)	Breaking of cell wall; damage to root cell vacuoles	Mirzajani et al. (2013)
AgNPs – 35–40 nm	Common wheat (*Triticum aestivum*)	Optimum growth promotion; improved shoot parameters and increased root nodulation observed	Pallavi et al. (2016)

9.3.3 NANOPESTICIDES INDUCED TOXICITY IN ANIMALS

Diversity of organisms residing in the soil aquatic and terrestrial environment is enormous and varies according to nature and availability of resources. One such very important organism is Earthworm. Earthworms are important creatures in the soil environment and are referred to as "ecosystem engineers." Due to their burrowing activities, they are recognized for enhancing soil microbial biomass, raising denitrifying microbe abundance and total N_2O emissions, and changing the distribution of the soil microbial community. The organic components found in earthworm mucus, which is high in carbohydrates and protein-like substances, promote the formation and activity of a large microbial community (Heggelund et al., 2013). Nanopesticides are found to alter the gut microbiota and development of earthworms, which can have an indirect effect on how the soil microbial community functions and, we assume that normal activities of earthworm including the digestion of soil organic matter in the stomach and subsequently the release of epidermal mucus may be crucial in controlling the toxicity of nanoparticles to the structure and function of the soil bacterial community. Alterations or changes in the microbial community structure may be a sign of adjustments in the function and metabolism of the microbial community in soil ecosystems.

$Cu(OH)_2$ nanopesticide (CNPE)- This copper-based nanopesticide may be used on trees, fruits, and vegetables in organic farming. Zebrafish (*Danio rerio*) serves as a replica model organism in nanotoxicity tests for aquatic animals as they have a number of traits, including high fertility, quick development of embryo, tiny size, and simplicity in culture. Genetic makeup of zebrafish is also related to that of humans. The results showed the dependency on the increase in dosage concentration, it was observed that death embryo numbers increased within the first 72 hours after exposure.

In aquatic and soil conditions, cadmium (Cd) contamination has been treated with nano-hydroxyapatite (NHAp). It was determined that the compound NHAp-Cd causes enhanced lipid peroxidation, catalase, and peroxidase activity in *Daphnia magna*, indicating oxidative stress in this species (Aksakal and Arslan, 2019). NHAp is quickly absorbed by cells and has been shown to cause cell death by inducing inflammatory reactions and oxidative stress. Depending on its shape, this nanomaterial can induce apoptosis and even cause epithelial as well as endothelial damages (Santos et al., 2018; Guo et al., 2020). However, further studies assessing the safety of ingested NHAp in more complex animals are still required (Rao et al., 2019).

Zinc based nanoparticles that might be a part of nanopesticides cause toxicity to major aquatic organisms such as *Artemia salina, D. magna,* and *Mytilus edulis* in a concentration dependent manner. In earthworms (*Eisenia fetida, Xiphinema vuittenezi,* and *C. elegans* species), ZnNP are better uptaken than Zn in ionic form, but also more toxic and studies further show decreased reproduction rates, increased mortality, and even increase in the number of apoptotic cells (Heggelund et al., 2013; Danabas et al., 2020; Sávoly et al., 2016).

In another study, intranasal exposure to zinc oxide nanoparticles (ZnO NPs), a component of some nanopesticides, in *Rattus rattus* (rats) demonstrated the persistence of these nanoparticles in the liver and other organs. This accumulation led to altered expression of metabolic and antioxidant genes, particularly in the liver (Tang et al., 2008). Similarly, copper nanoparticles (CuNPs) showed high toxicity to three fish species: rainbow trout (*Oncorhynchus mykiss*), fathead minnow (*Pimephales promelas*), and zebrafish (*Danio rerio*). The lowest-observed-effect concentrations (LOEC) were below the environmental concentrations predicted for CuNPs (Song et al., 2015). Additionally, at temperatures above 26°C, these nanoparticles tend to accumulate or aggregate, which can further increase toxicity to aquatic organisms.

9.3.4 NANOPESTICIDES INDUCED TOXICITY IN OTHER NON-TARGETED ORGANISMS

Although nanopesticides are used far more often than traditional pesticides, roughly 60% of nanopesticides still fail to kill the organisms they are intended to kill. In addition, polar solvents like acetone, methanol and others including alcohols, benzenes, and ketones, which have significant acute toxicity and are quickly absorbed into cropland and groundwater, make up the majority of the emulsifiable concentrates and microemulsions employed in nanopesticides. Polar solvents for instance acetone that is released into the environment as a carrier or by-product are difficult to break down and will build up there. This might lead to widespread acute animal poisoning and perhaps death. Knowing that nanopesticides are smaller, more capable of entering cells, and more accessible to the targeted microorganisms than traditional pesticides, the following questions must be raised: nanopesticides dispersed throughout the environment have a higher propensity than ordinary pesticides to enter unintended targets and cause unforeseen damage to those targets. Nano-TiO_2 insecticides can destroy both dangerous and helpful microorganisms throughout the agricultural production process.

Cadmium sulfide (CdS) nanopesticides produced by green synthesis have the likely possibility to kill a significant number of aquatic creatures that are not specifically targeted, including river crabs. Without a doubt, dangerous nanopesticides still present in edible plants and animals have a high likelihood of entering human bodies through the food chain and endangering human health. Assessments should consider potential effects across generations as well as the complete lifespan in order to improve the empirical data that can be used in the risk-assessment processes. The necessity to comprehend how aged/weathered nanomaterials affect ecological processes is highlighted by the discovery that soil ageing also causes changes in nanomaterial-induced toxicity in soil animals.

It remains a significant problem to apply the limited scientific understanding of the behaviour, destiny, bioavailability, and impacts of nanomaterials in the environment to regulation and policy.

9.4 TOXICITY AND ENVIRONMENTAL IMPACTS

More analysis and study are needed to address concerns regarding the toxicity and environmental effect of formulations based on nanoscales. The projected ambient concentration of nanoformulations in soil, surface water, groundwater, as well as any possible effects on nontarget animals, are still little understood. In addition to the properties of the nanoformulations detailed, environmental destiny is further influenced by variables in the environment, such as ionic strength, pH and co-molecules that are dissolved in the medium.

It has been thought that because nanomaterial-enabled insecticides, fertilizers or pesticides for that matter are being tuned for longer and sustained release, they would be more effective at minimal application rates and possibly could have less of an impact on the environment. The sensitivity (which is the limit of detection in terms of particle numbers) and specificity (which is the limit of detection in terms of particle numbers) necessary for realistic exposure concentrations at environmental level are both absent from many standard nanocharacterization technologies (e.g., DLS, TEM). The interactions of most nanoparticles in diverse environmental systems, including their harmful mechanisms, biological reactions of non-target organisms, bioavailability, assimilation in beneficial organisms and creatures and subsequent fate, are the critical problems or issues that have to be addressed at the earliest.

9.5 ASSESSMENTS AND REGULATORY AGENCIES

In continuation of the topic, various government and independent regulatory agencies, such as the Scientific Advisory Panel under the Federal Insecticide, Fungicide, and Rodenticide Act (FIFRA) of the United States, in a recent meeting debated and discussed the effectiveness of current hazard and exposure evaluation procedures, reaching the following conclusions:

1. The potential risks of pesticides containing engineered nanoparticles (ENPs) to humans and the environment may differ from those of conventional pesticides.
2. The current environmental fate and effects models are notably inadequate for predicting the behavior of nanopesticides in the long term.
3. Additional metrics and parameters are required to improve knowledge of the environmental fate and impact processes associated with nanopesticides.

The novel properties of a nanopesticide must also be taken into account in the assessment process under European Union (EU) Regulation 1107/2009, which calls for careful consideration of interactions among active ingredients (AI), safeners (chemicals used in combination with pesticides to make them safer and more targeted), synergists, and co-formulants when evaluating a plant protection product (Van den Brink et al., 2018).

Research on occupational and environmental safety by various regulatory authorities has lagged far behind developments in nanotechnology. For instance, to determine the potential harmful effects of nanotechnology on worker health, the National Institute for Occupational Safety and Health (NIOSH) established a Nanotechnology Research Center in 2004. This Center also supports businesses in developing and implementing strategies to protect employees from any recognized negative health impacts associated with working with nanomaterials.

Two main factors to be considered when evaluating the risks of using nanoagrochemicals to humans are low environmental consequences (on water resources and residues in food and dairy products) and worker safety, as well as social, ethical, and legal implications of nanotechnology (Kookana et al., 2014). The Global Horizon Scanning Project (GHSP) is an innovative project that aims to identify critical global environmental quality research needs in this context. This project has identified 40 research priorities for North America, 22 for Europe, and 20 significant topics regarding environmental pollution and contamination from Latin America (Fairbrother et al., 2019). These inquiries focus mainly on the fundamental issues of which formulations and chemicals we should be most concerned about in terms of their negative effects on the environment, as well as the sustainability of formulations and chemicals through nanotechnology.

It is widely recognized that the regulatory framework overseeing nanoagrochemical development has a significant impact on their introduction to the market. The National Nanotechnology Initiative (NNI) in the USA coordinates government and academic research projects on the effects of nanotechnologies on ecology and the environment. Additionally, the United States Food and Drug Administration (FDA) has already issued guidelines for using nanomaterials, particularly nanofertilizers, in animal feed. The information gathered so far suggests that food products containing nanoparticles currently on the market are likely safe for consumption, though further research is needed. In fact, to ensure the safe use of such agrochemicals, it is necessary to better understand their modes of action in line with standardized assessment methodologies before nanoagrochemicals can be used on a large scale.

Shifting focus to the Indian scenario, products that contain embedded or encapsulated nanoparticles are referred to as nano-agri products (NAPs). According to guidelines from agencies working under national regulatory provisions, these products are regulated by the Central Insecticide Board and Registration Committee (CIB&RC), the Food Safety and Standards Authority of India (FSSAI), the Fertilizer (Control) Order (FCO), and the Bureau of Indian Standards (BIS).

9.6 CONCLUSION

When comparing the advantages of nanoparticles to conventional pesticides, nanopesticides present a superior alternative. The environmental fate, behavior, and impact of nanopesticides may vary significantly depending on their source, shape, and size. However, the indiscriminate use of nanopesticides fosters conditions in which regulatory intervention is essential. Environmental risk assessments of nanopesticides also provide metrics on end-level exposure in the environment. Recommendations for creating a regulatory framework for nanopesticides are necessary. These recommendations do not need to be exhaustive; instead, they should serve as a general guide covering various components—such as characterization, environmental fate, exposure, and adverse effects—to ensure an appropriate regulatory response for nanopesticides.

KEYWORDS

- **active ingredients**
- **conventional pesticides**
- **forensic biology**
- **nanocapsules**
- **nanoemulsions**
- **nanoparticles**
- **nanopesticides**
- **toxicity**

REFERENCES

Abdelkhalek, A., Yassin, Y., Abdel-Megeed, A., Abd-Elsalam, K. A., Moawad, H., & Behiry, S. I. (2022). *Rhizobium leguminosarum* bv. *viciae*-mediated silver nanoparticles for controlling bean yellow mosaic virus (BYMV) infection in faba bean plants. *Plants*, *12*(1), 45. https://doi.org/10.3390/plants12010045.

Aksakal, F. I., & Arslan, H. (2019). Detoxification and reproductive system-related gene expression following exposure to $Cu(OH)_2$ nanopesticide in water flea (*Daphnia magna* Straus 1820). *Environmental Science and Pollution Research*, *27*(6), 6103–6111. https://doi.org/10.1007/s11356-019-07414-x.

Ameen, F., Alsamhary, K., Alabdullatif, J. A., & ALNadhari, S. (2021). A review on metal-based nanoparticles and their toxicity to beneficial soil bacteria and fungi. *Ecotoxicology and Environmental Safety*, *213*, 112027. https://doi.org/10.1016/j.ecoenv.2021.112027.

Arnaout, C. L., & Gunsch, C. K. (2012). Impacts of silver nanoparticle coating on the nitrification potential of *Nitrosomonas europaea*. *Environmental Science & Technology*, *46*(10), 5387–5395. https://doi.org/10.1021/es204540z.

Berahmand, A. A., Ghafariyan Panahi, A., Sahabi, H., Feizi, H., Rezvani Moghaddam, P., Shahtahmassebi, N., Fotovat, A., Karimpour, H., & Gallehgir, O. (2012). Effects of silver nanoparticles and magnetic field on growth of fodder maize (*Zea mays* L.). *Biological Trace Element Research*, *149*(3), 419–424. https://doi.org/10.1007/s12011-012-9434-5.

Danabas, D., Ates, M., Ertit Tastan, B., Cicek Cimen, I. C., Unal, I., Aksu, O., & Kutlu, B. (2019). Effects of Zn and ZnO nanoparticles on *Artemia salina* and *Daphnia magna* organisms: Toxicity, accumulation and elimination. *Science of the Total Environment*, *711*, 134869. https://doi.org/10.1016/j.scitotenv.2019.134869.

Deng, R., Lin, D., Zhu, L., Majumdar, S., White, J. C., Gardea-Torresdey, J. L., & Xing, B. (2017). Nanoparticle interactions with co-existing contaminants: Joint toxicity, bioaccumulation, and risk. *Nanotoxicology*, *11*(5), 591–612. https://doi.org/10.1080/17435390.2017.1343404.

Dimkpa, C. O., McLean, J. E., Martineau, N., Britt, D. W., Haverkamp, R., & Anderson, A. J. (2013). Silver nanoparticles disrupt wheat (*Triticum aestivum* L.) growth in a sand matrix. *Environmental Science & Technology*, *47*(2), 1082–1090. https://doi.org/10.1021/es302973y.

European Commission (EC). (2009). Regulation (EC) No. 1107/2009 of the European Parliament and the Council of 21 October 2009 concerning the placing of plant protection products on the market and repealing Council Directives 79/117/EEC and 91/414/EEC. *Official Journal of the European Union*, *309*, 1–50.

European Food Safety Authority (EFSA). (2013). Draft guidance document on tiered risk assessment for plant protection products for aquatic organisms in edge-of-field surface waters. *EFSA Journal*, *11*(7), 3290. https://doi.org/10.2903/j.efsa.2013.3290.

Fairbrother, A., Muir, D., Solomon, K. R., Ankley, G. T., Rudd, M. A., Boxall, A. B. A., Apell, J. N., Armbrust, K. L., Blalock, B. J., Bowman, S. R., Campbell, L. M., Cobb, G. P., Connors, K. A., Dreier, D. A., Evans, M. S., Henry, C. J., Hoke, R. A., Houde, M., Klaine, S. J., Klaper, R. D., & Brooks, B. W. (2019). Toward sustainable environmental quality: Priority research questions for North America. *Environmental Toxicology and Chemistry*, *38*(8), 1606–1624. https://doi.org/10.1002/etc.4502.

FIFRA-SAP. (2009). Evaluation of the hazard and exposure associated with nanosilver and other nanometal pesticide products. U.S. Federal Insecticide, Fungicide, and Rodenticide Act (FIFRA) Scientific Advisory Panel (SAP). http://www.epa.gov/scipoly/sap/meetings/2009/november/110309ameetingminutes.pdf (accessed on 3 October 2024).

Guo, Z., Luo, Y., Zhang, P., Chetwynd, A. J., Xie, H. Q., Abdolahpur Monikh, F., Tao, W., Xie, C., Liu, Y., Xu, L., Zhang, Z., Valsami-Jones, E., Lynch, I., & Zhao, B. (2020). Deciphering the particle specific effects on metabolism in rat liver and plasma from ZnO nanoparticles versus ionic Zn exposure. *Environment International*, *136*, 105437. https://doi.org/10.1016/j.envint.2019.105437.

Heggelund, L. R., Diez-Ortiz, M., Lofts, S., Lahive, E., Jurkschat, K., Wojnarowicz, J., Cedergreen, N., Spurgeon, D., & Svendsen, C. (2013). Soil pH effects on the comparative toxicity of dissolved zinc, non-nano, and nano ZnO to the earthworm *Eisenia fetida*. *Nanotoxicology*, *8*(5), 559–572. https://doi.org/10.3109/17435390.2013.809808.

Huang, B., Chang, F., Shen, Y., Qian, K., Wang, Y., Sun, C., Zhao, X., Cui, B., Gao, F., Zeng, Z., & Cui, H. (2018). Advances in targeted pesticides with environmentally responsive controlled release by nanotechnology. *Nanomaterials*, *8*(2), 102. https://doi.org/10.3390/nano8020102.

Kah, M., Kookana, R. S., Gogos, A., & Bucheli, T. D. (2018). A critical evaluation of nanopesticides and nanofertilizers against their conventional analogues. *Nature Nanotechnology*, *13*(8), 677–684. https://doi.org/10.1038/s41565-018-0131-1.

Kah, M., Weniger, A. K., & Hofmann, T. (2016). Impacts of (nano)formulations on the fate of an insecticide in soil and consequences for environmental exposure assessment. *Environmental Science & Technology*, *50*(20), 10960–10967. https://doi.org/10.1021/acs.est.6b02477.

Kookana, R. S., Boxall, A. B. A., Reeves, P. T., Ashauer, R., Beulke, S., Chaudhry, Q., Cornelis, G., Fernandes, T. F., Gan, J., Kah, M., Lynch, I., Ranville, J., Sinclair, C., Spurgeon, D., Tiede, K., & Van den Brink, P. J. (2014). Nanopesticides: Guiding principles for regulatory evaluation of environmental risks. *Journal of Agricultural and Food Chemistry*, *62*(19), 4227–4240. https://doi.org/10.1021/jf500232f.

Liu, Y., Li, S., Chen, Z., Megharaj, M., & Naidu, R. (2014). Influence of zero-valent iron nanoparticles on nitrate removal by *Paracoccus* sp. *Chemosphere*, *108*, 426–432. https://doi.org/10.1016/j.chemosphere.2014.02.045.

Mirzajani, F., Askari, H., Hamzelou, S., Farzaneh, M., & Ghassempour, A. (2013). Effect of silver nanoparticles on *Oryza sativa* L. and its rhizosphere bacteria. *Ecotoxicology and Environmental Safety*, *88*, 48–54. https://doi.org/10.1016/j.ecoenv.2012.10.018.

Nanopesticides Market Size, Growth, Trends, Report 2022-2030. (n.d.). *Precedence Research*. Retrieved from: https://www.precedenceresearch.com/nanopesticides-market (accessed on 3 October 2024).

Pallavi, Mehta, C. M., Srivastava, R., Arora, S., & Sharma, A. K. (2016). Impact assessment of silver nanoparticles on plant growth and soil bacterial diversity. *3 Biotech*, *6*(2), 1–11. https://doi.org/10.1007/s13205-016-0567-7.

Pandey, S., Giri, K., Kumar, R., Mishra, G., & Raja Rishi, R. (2016). Nanopesticides: Opportunities in crop protection and associated environmental risks. *Proceedings of the National Academy of Sciences, India Section B: Biological Sciences*, *88*(4), 1287–1308. https://doi.org/10.1007/s40011-016-0791-2.

Pradhan, S., & Mailapalli, D. R. (2017). Interaction of engineered nanoparticles with the agri-environment. *Journal of Agricultural and Food Chemistry*, *65*(38), 8279–8294. https://doi.org/10.1021/acs.jafc.7b02528.

Rao, C. Y., Sun, X. Y., & Ouyang, J. M. (2019). Effects of physical properties of nano-sized hydroxyapatite crystals on cellular toxicity in renal epithelial cells. *Materials Science and Engineering: C*, *103*, 109807. https://doi.org/10.1016/j.msec.2019.109807.

Rastogi, A., Zivcak, M., Sytar, O., Kalaji, H. M., He, X., Mbarki, S., & Brestic, M. (2017). Impact of metal and metal oxide nanoparticles on plants: A critical review. *Frontiers in Chemistry*, *5*, 1–16. https://doi.org/10.3389/fchem.2017.00078.

Santos, C., Turiel, S., Sousa Gomes, P., Costa, E., Santos-Silva, A., Quadros, P., Duarte, J., Battistuzzo, S., & Fernandes, M. H. (2018). Vascular biosafety of commercial hydroxyapatite particles: Discrepancy between blood compatibility assays and endothelial cell behavior. *Journal of Nanobiotechnology*, *16*(1), 1–12. https://doi.org/10.1186/s12951-018-0357-y.

Sávoly, Z., Hrács, K., Pemmer, B., Streli, C., Záray, G., & Nagy, P. I. (2016). Uptake and toxicity of nano-ZnO in the plant-feeding nematode, *Xiphinema vuittenezi*: The role of dissolved zinc and nanoparticle-specific effects. *Environmental Science and Pollution Research*, *23*(10), 9669–9678. https://doi.org/10.1007/s11356-015-5983-4.

Simonin, M., Colman, B. P., Tang, W., Judy, J. D., Anderson, S. M., Bergemann, C. M., Rocca, J. D., Unrine, J. M., Cassar, N., & Bernhardt, E. S. (2018). Plant and microbial responses to repeated $Cu(OH)_2$ nanopesticide exposures under different fertilization levels in an agro-ecosystem. *Frontiers in Microbiology*, *9*, 1–13. https://doi.org/10.3389/fmicb.2018.01769.

Song, L., Vijver, M. G., Peijnenburg, W. J. G. M., Galloway, T. S., & Tyler, C. R. (2015). A comparative analysis on the *in vivo* toxicity of copper nanoparticles in three species of freshwater fish. *Chemosphere*, *139*, 181–189. https://doi.org/10.1016/j.chemosphere.2015.06.021.

Tang, J., Xiong, L., Wang, S., Wang, J., Liu, L., Li, J., Wan, Z., & Xi, T. (2008). Influence of silver nanoparticles on neurons and blood-brain barrier via subcutaneous injection in rats. *Applied Surface Science*, *255*(2), 502–504. https://doi.org/10.1016/j.apsusc.2008.06.058.

Van den Brink, P. J., Boxall, A. B. A., Maltby, L., Brooks, B. W., Rudd, M. A., Backhaus, T., Spurgeon, D., Verougstraete, V., Ajao, C., Ankley, G. T., Apitz, S. E., Arnold, K., Brodin, T., Cañedo-Argüelles, M., Chapman, J., Corrales, J., Coutellec, M. A., Fernandes, T. F., Fick, J., & Ford, A. T. (2018). Toward sustainable environmental quality: Priority research questions for Europe. *Environmental Toxicology and Chemistry*, *37*(9), 2281–2295. https://doi.org/10.1002/etc.4205.

VandeVoort, A., & Arai, Y. (2018). Macroscopic observation of soil nitrification kinetics impacted by copper nanoparticles: Implications for micronutrient nanofertilizer. *Nanomaterials*, *8*(11), 1–16. https://doi.org/10.3390/nano8110927.

Wu, F., Falfushynska, H., Dellwig, O., Piontkivska, H., & Sokolova, I. M. (2020). Interactive effects of salinity variation and exposure to ZnO nanoparticles on the innate immune system of a sentinel marine bivalve, *Mytilus edulis*. *Science of the Total Environment*, *712*, 136473. https://doi.org/10.1016/j.scitotenv.2019.136473.

Zhang, L., Wu, L., Si, Y., & Shu, K. (2018). Size-dependent cytotoxicity of silver nanoparticles to *Azotobacter vinelandii*: Growth inhibition, cell injury, oxidative stress and internalization. *PLOS ONE*, *13*(12), 1–19. https://doi.org/10.1371/journal.pone.0209020.

Zhang, X., Xu, Z., Wu, M., Qian, X., Lin, D., Zhang, H., Tang, J., Zeng, T., Yao, W., Filser, J., Li, L., & Sharma, V. K. (2019). Potential environmental risks of nanopesticides: Application of $Cu(OH)_2$ nanopesticides to soil mitigates the degradation of neonicotinoid thiacloprid. *Environment International*, *129*, 42–50. https://doi.org/10.1016/j.envint.2019.05.022.

Zhao, L., Huang, Y., Adeleye, A. S., & Keller, A. A. (2017). Metabolomics reveals $Cu(OH)_2$ nanopesticide-activated antioxidative pathways and decreased beneficial antioxidants in spinach leaves. *Environmental Science & Technology*, *51*(17), 10184–10194. https://doi.org/10.1021/acs.est.7b02163.

Zulfiqar, F., Navarro, M., Ashraf, M., Akram, N. A., & Munné-Bosch, S. (2019). Nanofertilizer use for sustainable agriculture: Advantages and limitations. *Plant Science*, *289*, 110270. https://doi.org/10.1016/j.plantsci.2019.110270.

CHAPTER 10

Impact of Nanopesticides on Non-Targeted Organisms

TUSHAR R. AHIRE,[1,2] SUKANYA SACHDEVA,[1,2] DIVYA KOTTADIYIL,[3] and PERUMAL SIVAPERUMAL[1]

[1]*Chemical Sciences Division, ICMR–National Institute of Occupational Health, Ahmedabad, Gujarat, India*

[2]*Department of Biochemistry and Forensic Science, Gujarat University, Ahmedabad, Gujarat, India*

[3]*School of Behavioural and Forensic Sciences, JSPM University, Pune, India*

ABSTRACT

Nanomaterials have been widely employed globally in agricultural practices to protect crops from various pests and insects. This promising technique, due to its nanoscale size, enables efficient action against harmful organisms, with reduced quantities, rapid action of nanopesticides, and minimal harm to crops compared to traditional pesticides. However, while sprayed on crops to reduce pest activity, these nanopesticides simultaneously accumulate in soil, air, and water bodies, inadvertently affecting non-target organisms. Therefore, monitoring the hazards caused is essential to maintain environmental balance.

The use of modified pesticides is not yet widely adopted in developing countries like India, but it may gain popularity in the near future as Indian farmers become aware of the advantages of nanopesticides in boosting crop yield. However, nanopesticides can have unintended impacts on other

Nanopesticides: A Perspective from Welfare to Risk Assessment.
Garima Awasthi, Payal Lodha, and Gajendra Pal Singh (Eds.)

DOI: 10.1201/9781003605874-10

organisms, such as beneficial soil and crop organisms, airborne organisms, aquatic creatures, plants, and even humans. This indiscriminate impact disrupts the balance of natural ecosystems, affecting untargeted animals and ultimately disturbing the cycle of nature and humans.

Biodiversity is increasingly threatened by the use of nanopesticides. Scientists are deeply concerned about the chronic and acute effects on biodiversity and the food chain. The food chain, comprising numerous organisms, is disrupted by the aftereffects of nanopesticides. Understanding the hazards associated with nanopesticides and finding ways to mitigate their presence in the ecosystem is now a critical step forward.

10.1 INTRODUCTION

Agriculture is regarded as one of the most significant and stable sectors because it adds to the nation's economic ability to ensure food security. Due to factors including population growth, climate change, polluted environment, and increased yet depleting natural resource demands, food production and its distribution are posing a great deal of stress at the global level (Raj et al., 2021). The quantity and quality of food production are likely to be jeopardized as the world population continues to grow at an exponential rate, with the population anticipated to expand by roughly 48% by the year 2050. The world's agricultural regions deal with a variety of challenges, including decreases in soil organic matter, stagnant crop yields, dwindling arable land, water scarcity, multi-nutrient deficiencies, inadequacies, and a labor shortfall brought by migration from farming (Raliya et al., 2018).

India has been an agricultural country, and its economy thrives on it. A significant source of worry is the enormous number of insects and pests that harm crops, resulting in crop rotting. The major threat to agriculture and grain crops that are being harvested comes from insects and pests, which have a significant impact on the goods' quality, quantity, and market value (Taghizadeh and Mohammadkhani, 2017). The currently preferred approach to address this problem is to use agricultural chemistry. Nanotechnology offers tremendous potential, with chemical pesticides as the primary means of pest control. The use of nanostructures as agricultural chemicals (pesticides or fertilizers) with the aim of increasing the production of crops is constantly being addressed due to their distinctive physicochemical characteristics (Raliya et al., 2018). The development of these novel plant protection products, known as nanopesticides, which use nano-sized materials to bring innovative technological and environmental benefits, has occurred over the past decade (Gomes et al., 2019; Grillo et al., 2021; Sułowicz et al., 2023; Peixoto et al., 2021; Fojtová et al., 2019).

The rapid development of nanomaterials is responsible for state-of-the-art transformation in science and technology, enabling their use in a variety of industrial, agricultural, and manufacturing sectors and potentially transforming a wide range of disciplines. The application of nanopesticides against pests, which includes fungi, weeds, insects, and nanofertilizers as plant "growth-promoting" agents, has had a significant impact on plant science and the agricultural industry (Grillo et al., 2021). As a cutting-edge and alluring topic, nanotechnology offers new opportunities for developing nanoscale compounds, formulations, and their method of delivery, all encompassed under one heading "nanofertilizers." It is described as the use of nanotechnology to protect crops from various insects and pests. The interactions between nanoparticles (NPs) and insects, developing formulation with target analytes and incorporating them into nanopesticides, using NPs as active pesticide agents or as a carrier for pesticide delivery are all part of the study of nanopesticides (Yasur and Rani, 2013; Benelli and Lukehart, 2017; Benelli, 2016). Concerns regarding environmental build-up and, thus, possible eco-toxicity and human health risk arise from the extensive usage of nanomaterials, particularly when directly introduced into the environment via agricultural practices (Grillo et al., 2021). Nanotechnology manipulates molecules and atoms in order to formulate devices, systems, and materials (Chugh and Jahangir Dar, 2021). In the environment, concentrations of nanomaterials were found to be as low as of the order of ng/L, despite the fact that there is a dearth of data on these concentrations globally. There is currently no clear-cut definition of what constitutes a nanopesticide (Grillo et al., 2021). Some researchers have developed formulations with particle sizes of up to 1000 nm, which is contrary to the standard average range of 1–100 nm given in ISO/TS 80004-2:2015. Though the targets and modes of action of both pesticides and biocides differ significantly and are illustrated in separate regulatory regimes, yet some papers have referred to their nano-forms as nanopesticides.

With the aim to improve distribution and efficacy, "nanopesticides" and "nanofertilizers" were synthesized via formulating active ingredients in conjunction with nanotechnology. Alongside efficacy and target delivery due to nanoscale formulations, there comes the enhanced dispersion stability with controlled release of active analyte at the target site, which is one of the crucial aspects of good agricultural practices (Grillo et al., 2021).

Nanopesticides can easily disperse into the environment (air, water, and soil) via both intentional and unintentional use, similar to traditional pesticides. These are employed at low concentrations, and due to further dispersion into different components of the environment, calculating their concentration and thereby assessing the risk associated with their usage becomes a challenging task (Grillo et al., 2021). The environment and human health may be at risk from the use of engineered nanoscale materials in farming, food production, and water treatment (Chugh and Jahangir Dar, 2021). Since nanopesticides are newer on the market, their impact on the ecosystem is yet to be determined. Hence, studies focused on the risk assessment and impact of nanopesticides must be initiated (Sułowicz et al., 2023).

TABLE 10.1 Types of Nanopesticides with Examples

Sl. No.	Type of Nanopesticide	Subtypes	Examples
1	Organic NPs	Lipid-based nanopesticides	Liposomes, plant oil-based emulsions
2	Protein-based nanopesticides	Virus capsids	*Bacillus thuringiensis*, biodegradable corn-derived zein
3	Polymer-based nanopesticides	Naturally occurring, Synthetic	Chitosan, alginate, starch, polycaprolactone (PCL), polyethylene glycol (PEG), polylactic acid (PLA)
4	Inorganic nanopesticides	Metals	Silica, titanium dioxide, silver, copper, aluminum, carbon nanotubes
5	Organic-inorganic nanopesticides	Hybrids	Porous hollow nanoparticles of silica, nanospheres and nanocapsules of silica, clay minerals (cationic, anionic), MWCNT-g-PCA

10.2 NEGATIVE EFFECTS OF NANOPESTICIDES ON UNINTENDED ORGANISMS

When nanoparticles (NPs) are incorporated into agriculture, they can conduct agricultural operations efficiently but may lead to unintended adverse health impacts. In the previous section, it was discussed that the dispersion of nanomaterials onto plants could lead to bioaccumulation throughout the food chain. This bioaccumulation may pose potential risks to human and environmental health (Nisha Raj et al., 2021). For instance, silver nanoparticles (AgNPs), like other NPs, have shown potential harm to humans and other non-target organisms when exposure exceeds safe limits. In mammals, excessive AgNPs exposure may lead to neurotoxicity, inflammation, and other detrimental effects (Dangi and Kamra, 2021).

10.2.1 ENVIRONMENTAL AND HUMAN HEALTH

Although nanopesticides have a significantly higher utilization rate than standard pesticides, approximately 60% of nanopesticides still fail to affect their intended organisms (Xu et al., 2018; Kumar et al., 2019; Kah and Hofmann, 2014; Lo et al., 2008). Key questions regarding these modern pesticides concern whether they interact specifically with the environment, potentially affecting plants, the ecosystem, or disrupting the balance of the food chain. For instance, the accumulation of artificial nanomaterials in human tissues can cause adverse side effects, such as stress, endocrine disorders, infertility, and disruption of brain and muscle activity (Mailapalli, 2021). Additionally, exposure to aerosolized nanopesticides through inhalation poses health risks for humans and animals, with outcomes that vary depending on particle size, concentration, environmental conditions, exposure level, and the physicochemical characteristics of the particles (Raliya et al., 2018).

The microemulsions and emulsifiable concentrates used in nanopesticides often involve polar solvents like alcohols and ketones, which are acutely toxic and can readily contaminate cropland and groundwater (Alexander, 1985). Released into the environment, these polar solvents are challenging to degrade and may accumulate, potentially causing widespread acute toxicity in animals. Given the smaller size, greater cell penetration, and higher accessibility of nanopesticides compared to traditional pesticides, questions arise: Are nanopesticides more likely than conventional pesticides to infiltrate non-target organisms, potentially yielding unpredictable effects? For example, nano-TiO_2 insecticides can eliminate both harmful and beneficial bacteria in agricultural production, and green-synthesized CdS nanopesticides have shown toxic effects on numerous non-target aquatic species, including river crabs (Sun et al., 2019; Kalimuthu et al., 2017).

Research indicates that non-targeted effects on organisms cannot be completely ruled out, reinforcing the importance of designing safe nanopesticide formulations. For example, double-stranded RNA (dsRNA) in pesticide design shows reduced persistence in the environment—soil, water, and biota—thereby limiting exposure to non-target organisms (Lucia and Renato, 2019). However, nanopesticides present in edible plants and animals may enter the human food chain, potentially posing health risks. Toxicity can be measured using the minimum biocidal concentration (MBC), defined as the concentration (μg/mL) that completely inhibits microbial growth after a day's exposure. Mycorrhizal fungi, rhizobacteria, and cellulose- and lignin-degrading fungi, which support plant growth and soil fertility, may be negatively impacted by the antibacterial properties of NPs. Nano-fungicides have shown 10- to 100-fold reductions in growth inhibition of off-target microbial strains compared to traditional fungicides like captan (Sułowicz et al., 2023).

Nanoemulsions and nanodispersions can enhance the solubility of poorly water-soluble pesticides, improving bioavailability of active compounds and eliminating adjuvants that may harm non-target organisms. However, biosafety and chronic effects of nanopesticides on humans and the environment remain concerns, especially for agricultural workers regularly exposed. Approved nano-insecticides in Europe include synthetic amorphous silicon dioxide as stable aggregates >1 μm. When copper is used as a crop fungicide (US-EPA, 2009), it may exhibit unexpected biocidal and toxicological activity after nanoscale "biotransformation" in the environment (Iavicoli et al., 2017). Testing protocols for nanopesticides must account for unique properties due to nanoscale dimensions and their selective effects on aquatic, terrestrial, and aerial species (Grillo et al., 2021).

Environmental shifts in biodiversity may occur in response to nanopesticide exposure, with low concentrations of certain pesticides potentially supporting biodiversity but high concentrations leading to harmful effects, such as reduced microbial species richness and selective species impact. Stimulatory effects of low pesticide levels have been observed, such as imidacloprid's selective impact on beetle populations (Paederus fuscipes Curtis), and glyphosate's effects on specific algae and plants. Microbial species may thrive under low nanopesticide exposure, but higher concentrations can be lethal, potentially altering species richness and composition. Such shifts may impact the food chain, ultimately affecting biodiversity.

Toxicological studies emphasize the need for health and environmental risk assessments for low nanopesticide concentrations, which are more likely to occur environmentally than high concentrations (Agathokleous et al., 2019).

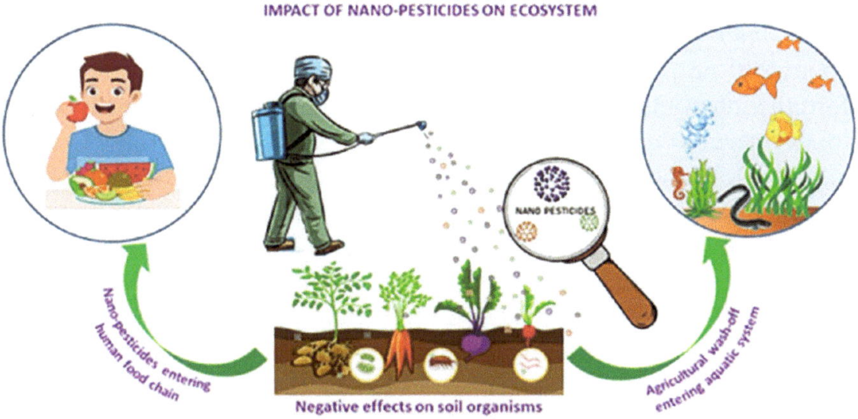

FIGURE 10.1 Impact of nanopesticides on ecosystem.

10.2.2 PLANT AND SOIL MICROORGANISMS

Pesticides based on nanotechnology may be more effective and/or safer than their traditional counterparts. However, little is known about how nanoparticles (NPs) impact plant physiology and metabolism when they interact with them, particularly in terms of their mode of action (Santiago et al., 2020). The behavior of nanomaterials and, consequently, how well they interact with plants and microbes, can be influenced by the characteristics of the soil. The chemical characteristics of NPs have been shown to be affected by components such as organic matter present in soil, ionic strength, pH, and phosphate content (Kalwani et al., 2022). When sprayed through the soil, especially when particles get mixed in the soil, engineered NPs have several negative effects: the direct exposure and localized concentration of the particles become significantly higher than indirect exposure (Raliya et al., 2018). Furthermore, the application of excessive dosages of nanofertilizers appears to have a considerable impact on the efficacy of nutrient absorption and utilization by plants. For instance, under the impact of nanofertilizers, the effective absorption of phosphorus, nitrogen, carbon, potassium, and sulfur has been lowered to 20–40% (Kalwani et al., 2022).

Atrazine, paraquat, and lipid NPs with diameters of around 290 nm and 300 nm, respectively, were much more harmful to *Caenorhabditis elegans*, a nematode, than their corresponding non-NP counterparts. In this investigation, exposure was brief and occurred through a solution rather than through soil (Grillo et al., 2021). Atrazine NPs (diameter: about 230 nm, EC10 ca. 180 mg/kg), pure atrazine (a.i.), and atrazine formulation commercially available (Gesaprim) were the least hazardous to *Enchytraeus crypticus* (EC10 ca. 100 mg/kg), respectively (Gomes et al., 2019). These NPs' toxicity may be linked to the increased ability of non-targeted organisms to internalize and possibly bioaccumulate them (Fojtová et al., 2019). When two earthworms (*Lumbricus terrestris* and *Eisenia fetida*) were tested with two bifenthrin formulations (Nano A and Nano B, obtained from Vive Crop Protection, size: n.a.), it was found that when compared to the non-nano version of bifenthrin, the accumulation was approximately 50% higher (Iavicoli et al., 2017). Commercial inorganic NPs, such as Kocide 3000® from DuPont Co.TM, with a diameter of more than 1000 nm, were more harmful to *Folsomia candida* than traditional formulations, such as $Cu(OH)_2$ solution. These differences were attributed to the active copper ingredient (Grillo et al., 2021).

The nematode *Caenorhabditis elegans* was shown to be non-toxic to neem oil-based protein NPs (diameter: ca. 288 nm), but the formulation without NPs interfered with the organism's pharyngeal pumping and expression of the GST-4 protein (Mohd et al., 2018). Even though the NPs were detected in the weevil, the same outcomes were observed when it was tested against *Acanthoscelides obtectus* (Pascoli et al., 2020). Moreover, non-target plant species have had the toxicity of NPs assessed. Seed germination and root length in *Lycopersicum esculentum*, *Zea mays*, and *Cucumis sativus* were unaffected by the permethrin NPs (diameter approximately 130 nm). When tested on off-target maize plants, an atrazine polymeric NP (diameter: about 250 nm) produced similar results (Science and John, 2012; Oliveira et al., 2015). Moreover, some research has employed a metabolomics method to comprehend how plants' metabolisms are altered when exposed to NPs (Grillo et al., 2021).

Because of their activity, invertebrates in agroecosystems may affect the bioavailability of pollutants in soils by enhancing soil oxygenation and organic matter input. Since invertebrate activity may, for example, boost soil aeration or enhance the microbial community, it is anticipated that microbial responses to contaminants will differ in the presence of invertebrates (Peixoto et al., 2021). Terrestrial isopods are an example of these organisms; they are widespread, regarded as strong bioindicators of soil quality, and certain species are synanthropic (Van Gestel et al., 2018). The abundance of heterotrophic bacteria in soil was negatively impacted after 90 days of exposure to the NPs' seasonal application rate, but $Cu(OH)_2$-ionic boosted the number of CFU/g soil. As the NPs demonstrated a much higher copper level than the control and the $Cu(OH)_2$-ionic after 90 days, these variations may be due to copper levels in pore water. Several mechanisms of action have been hypothesized, and numerous studies have shown the harmful effects of $Cu(OH)_2$-NPs on soil microbial communities or specific bacterial groups. The generation of reactive oxygen species (ROS), dissolved Cu^{2+}, $Cu(OH)_2$-specific effects, or oxidative stress can all cause these harmful consequences (i.e., superoxide or hydroxyl radicals) (Peixoto et al., 2021).

A lengthier exposure can drastically change the composition of the soil's microbial community (Kalwani et al., 2022). The structure and diversity of soil bacterial and fungal communities are impacted by long-term exposure to NPs, especially at the seasonal application dose, while microbial processes involved in the carbon cycle are impacted at both application rates. Similar to other copper-based nanomaterials, conventional insecticides ($CuSO_4$) and CuO NPs (e.g., Lufa 2.1; pH 7.6) clearly affected the microbial communities from calcareous soils (Peixoto et al., 2021). The effects of $Cu(OH)_2$ NPs on the soil microbial community were evaluated both with and without *Porcellionides pruinosus*, as well as in comparison to those obtained after exposure to $Cu(OH)_2$ ions. This soil invertebrate's presence heightens the realism of these observations since it represents the invertebrate community that is crucial to the decomposition process and closely linked with the soil microbial community (Bray et al., 2019; Zimmer and Topp, 2002; Peixoto et al., 2020). Peixoto et al. (2021) investigated the effects of commercial $Cu(OH)_2$ NPs on microbial communities' structure, abundance, and function over 90 days at single and seasonal agricultural application doses, both with and without the isopod *P. pruinosus*. The study found that soil microbial populations are affected by $Cu(OH)_2$ NPs, potentially altering their ecological function.

Bifenthrin concentration in the soil and soil pore water decreased during the absorption phase, correlating with an increase in bifenthrin concentration in earthworms. At the end of the absorption phase, 65–82% of the bifenthrin was extractable and associated with soil particles, 16–33% had dispersed or decomposed, around 1% was in the pore water, and approximately 1% had been ingested by earthworms (Mishra et al., 2020). Mishra et al. (2020) also investigated the effects of silver nanomaterials on specific bacterial groups—Actinobacteria, β-proteobacteria, α-proteobacteria, and Bacteroidetes—which are crucial for polysaccharide degradation, nitrogen cycling, and plant growth promotion. When exposed to silver NPs, bacterial abundance remained constant up to the 30th day; however, the quantity of α-proteobacteria then declined. Similarly, after exposure to nanomaterials, the abundance of actinobacterial species was reduced by day 30.

10.2.3 WATER AND AQUATIC ECOSYSTEM

The lethal indices of nanoparticle (NP) treatment versus bulk treatment showed significant differences ($P < 0.001$). An investigation on *Danio rerio* (Zebra Fish) exposed to high concentrations (> 6.4 mg/L) of permethrin nanoemulsion (PNE) demonstrated behavioral abnormalities, such as irregular swimming and lack of reflex. At higher PNE concentrations, zebra fish displayed more aberrant behaviors along with significant gill histology deformation due to the higher dosage of the insecticide pyrethroid. The acute toxicity tests revealed lethal concentration indices: LC10 = 3.9 mg/L, LC50 = 6.4 mg/L, and LC90 = 10.5 mg/L (Mishra et al., 2019). A risk assessment study by DeSilva and Pathiratne (2020) examined the histopathological and genotoxic effects of nano-TiO_2 (concentrations of 0.1, 0.5, and 1 mg/L) on *Oreochromis niloticus* (Nile Tilapia) over exposure times of 7, 14, and 21 days. According to the comet assay, nano-TiO_2 can cause erythrocytic DNA damage in a concentration-dependent manner. Increased organ damage indices revealed histological changes in the liver and intestine due to nano-TiO_2. Two severe histological abnormalities, necrosis of the intestinal mucosa and hepatic parenchyma, were noted (De Silva and Pathiratne, 2023). Examining the toxic effects of NP deltamethrin on the gill morphology of *L. rohita* (Rohu) showed considerable histological damage in the gills, which became more pronounced with NP deltamethrin treatment (Sahoo et al., 2017). Histological analysis of the gills from treated and control samples helped determine NPs' cytotoxic impact on zebra fish physiology, showing intact primary and secondary lamellae without deformation in control samples (Mishra et al., 2019).

Research on *V. pachynema* extracts underscores their efficiency as a reducing agent for CdS NP synthesis, with potential as mosquito repellents. The seaweed *V. pachynema* reduced CdS ions in water, aiding the development of metal NPs for pest management and showing antiplasmodial efficacy against CQ-r malaria parasites. CdS NPs and Cd ions in solution were found highly toxic to mud crabs, with Cd ions proving more hazardous than nano-CdS. *Scylla serrata* crabs exposed to sub-lethal doses of CdS NPs (4, 6, and 8 g/mL) showed changes in GST and AChE enzyme activity after 16 days, with increased GST activity and reduced AChE activity, indicating neurotoxicity. However, *S. serrata* could partially restore enzyme function, as depicted in the recovery pattern (Kalimuthu et al., 2017).

10.3 CONCLUSION

Nanopesticides, in particular, have shown significant promise for application in agriculture as a means of protecting crops from insects and pests. When compared to conventional pesticides, nanopesticides are more effective due to their smaller, faster-acting particles that can target harmful organisms more precisely while causing less damage to crops. However, nanopesticides can accumulate in soil, air, and water bodies, affecting non-target organisms and potentially causing unforeseen consequences. To maintain a healthy environmental balance, it is essential to monitor and assess the potential risks associated with nanopesticides. In developing nations like India, the use of modified pesticides remains uncommon; however, as awareness of the crop yield benefits associated with nanopesticides grows, their use may become more widespread in the future. It is crucial to recognize that nanopesticides may impact not only harmful organisms but also beneficial organisms for crops and soil, as well as airborne organisms, aquatic life, plants, and even humans. The indiscriminate use of nanopesticides could disrupt the delicate balance between nature and humans, emphasizing the need for responsible application. While nanopesticides hold the potential to revolutionize agriculture, it is vital to weigh their benefits against any potential risks to the environment and public health.

10.4 FUTURE PERSPECTIVE

Given that nanoparticles have proven considerable advantages over conventional pesticides, their use in agriculture is likely to increase in the future. By providing a more effective and environmentally benign means of protecting crops from pests and insects, nanopesticides have the potential to revolutionize agriculture. However, research on the long-term effects of nanopesticides is increasingly essential to address concerns about unintended environmental and health consequences. This research should focus on assessing the environmental risks related to their application, including impacts on non-target organisms and potential accumulation in soil, air, and water.

Furthermore, it is crucial to establish rules and guidelines to ensure nanopesticides are used responsibly and ethically. Formulating these guidelines should involve coordinated efforts among government agencies, scientists, and the agricultural industry to balance the need for improved crop yields with potential environmental and health effects. Overall, while nanoparticles hold tremendous promise for agricultural benefits, it remains essential to consistently monitor and evaluate associated risks and to develop reasonable rules and standards to guide their use responsibly.

KEYWORDS

- **aquatic ecosystem**
- **biodiversity**
- **inflammation**
- **neurotoxicity**
- **plant microorganisms**
- **soil microorganisms**
- **water ecosystem**

REFERENCES

Agathokleous, E., Feng, Z., Iavicoli, I., & Calabrese, E. J. (2019). Nanopesticides: A great challenge for biodiversity? The need for a broader perspective. *Nano Today*, *29*, 100808. https://doi.org/10.1016/j.nantod.2019.100808.

Alexander, M. (1985). Effect of Chemical Structure on Microbial Degradation of Substituted Benzenes, 410–413.

Benelli, G. (2016). Plant-mediated biosynthesis of nanoparticles as an emerging tool against mosquitoes of medical and veterinary importance: A review. *Parasitology Research*, *115*(1), 23–34. https://doi.org/10.1007/s00436-015-4800-9.

Benelli, G., & Lukehart, C. M. (2017). Special issue: Applications of green-synthesized nanoparticles in pharmacology, parasitology and entomology. *Journal of Cluster Science*, *28*(1), 1, 2. https://doi.org/10.1007/s10876-017-1165-5.

Bray, N., Kao-Kniffin, J., Frey, S. D., Fahey, T. J., & Wickings, K. (2019). Soil macroinvertebrate presence alters microbial community composition and activity in the rhizosphere. *Frontiers in Microbiology*, *10*, 256. https://doi.org/10.3389/fmicb.2019.00256.

Chaud, M., Souto, E. B., Zielinska, A., Severino, P., Batain, F., Oliveira-Junior, J., & Alves, T. (2021). Nanopesticides in agriculture: Benefits and challenge in agricultural productivity, toxicological risks to human health and environment. *Toxics*, *9*(6), 131. https://doi.org/10.3390/toxics9060131.

Chugh, P., & Jahangir Dar, S. (2021). Nanotechnology: Applications, opportunities, and constraints in agriculture. *Nanotechnology*, 125–148. https://doi.org/10.1201/9781003120261-4.

Dangi, K., & Kamra, A. (2021). Efficient & eco-friendly smart nanopesticides: Emerging prospects for agriculture. *Materials Today: Proceedings*, *45*, 3819–3824. https://doi.org/10.1016/j.matpr.2021.03.211.

De Silva, W. A. P. M., & Pathiratne, A. (2023). Nano-titanium dioxide induced genotoxicity and histological lesions in a tropical fish model, Nile tilapia (*Oreochromis niloticus*). *Environmental Toxicology and Pharmacology*, *98*, 104043. https://doi.org/10.1016/j.etap.2023.104043.

Del Prado-Audelo, M. L., Bernal-Chávez, S. A., Gutiérrez-Ruíz, S. C., Hernández-Parra, H., Kerdan, I. G., Reyna-González, J. M., Sharifi-Rad, J., & Leyva-Gómez, G. (2022). Stability phenomena associated with the development of polymer-based nanopesticides. *Oxidative Medicine and Cellular Longevity*, *2022*, 5766199. https://doi.org/10.1155/2022/5766199.

Fang, X., Zhang, Q., Yu, B., Wang, B., Yang, G., Wang, B., & Hua, Z. (2023). Plant-oil based polymeric emulsions as adhesive nanocarriers for enhancing the efficacy of nanopesticides. *Industrial Crops and Products*, *192*, 116020. https://doi.org/10.1016/j.indcrop.2022.116020.

Fojtová, D., et al. (2019). Nanoformulations can significantly affect pesticide degradation and uptake by earthworms and plants. *Environmental Chemistry*, *16*(6), 470–481. https://doi.org/10.1071/EN19057.

Gomes, S. I. L., Scott-Fordsmand, J. J., Campos, E. V. R., Grillo, R., Fraceto, L. F., & Amorim, M. J. B. (2019). On the safety of nanoformulations to non-target soil invertebrates-an atrazine case study. *Environmental Science: Nano*, *6*(6), 1950–1958. https://doi.org/10.1039/c9en00242a.

Grillo, R., Fraceto, L. F., Amorim, M. J. B., Scott-Fordsmand, J. J., Schoonjans, R., & Chaudhry, Q. (2021). Ecotoxicological and regulatory aspects of environmental sustainability of nanopesticides. *Journal of Hazardous Materials*, *404*, 124148. https://doi.org/10.1016/j.jhazmat.2020.124148.

Iavicoli, I., Leso, V., Beezhold, D. H., & Shvedova, A. A. (2017). Nanotechnology in agriculture: Opportunities, toxicological implications, and occupational risks. *Toxicology and Applied Pharmacology*, *329*, 96–111. https://doi.org/10.1016/j.taap.2017.05.025.
Jacques, M. T., Oliveira, J. L., Campos, E. V. R., Fraceto, L. F., & Ávila, D. S. (2017). Safety assessment of nanopesticides using the roundworm *Caenorhabditis elegans*. *Ecotoxicology and Environmental Safety*, *139*, 245–253. https://doi.org/10.1016/j.ecoenv.2017.01.045.
Kah, M., & Hofmann, T. (2014). Nanopesticide research: Current trends and future priorities. *Environment International*, *63*, 224–235. https://doi.org/10.1016/j.envint.2013.11.015.
Kalimuthu, K., Malathi, A., & Kumari, S. (2017). Green-synthesized CdS nanopesticides: Toxicity on young instars of malaria vectors and impact on enzymatic activities of the non-target mud crab *Scylla serrata*. *Aquatic Toxicology*, *185*, 165–173. https://doi.org/10.1016/j.aquatox.2017.04.015.
Kalwani, M., Chakdar, H., Srivastava, A., Pabbi, S., & Shukla, P. (2022). Effects of nanofertilizers on soil and plant-associated microbial communities: Emerging trends and perspectives. *Chemosphere*, *287*, 132107. https://doi.org/10.1016/j.chemosphere.2021.132107.
Kumar, S., Nehra, M., Dilbaghi, N., Marrazza, G., Aly, A., & Kim, K. (2019). Nano-based smart pesticide formulations: Emerging opportunities for agriculture. *Journal of Controlled Release*, *294*, 131–153. https://doi.org/10.1016/j.jconrel.2018.12.012.
Lo, E., Martı, E., Arias-este, M., Mejuto, J., & Garcı, L. (2008). The mobility and degradation of pesticides in soils and the pollution of groundwater resources, *123*, 247–260. https://doi.org/10.1016/j.agee.2007.07.011.
Lucia, V., & Renato, S. S. D. C. (2019). *Nanopesticides*. Springer.
Mailapalli, D. R. (2021). Nanopesticides for pest control. In *Nanopesticides for Pest Control* (Chapter 2).
Mishra, P., Choudhary, S., Gupta, V. K., & Kumar, S. (2019). Enhanced mosquitocidal efficacy of colloidal dispersion of pyrethroid nanometric emulsion with benignity towards non-target species. *Ecotoxicology and Environmental Safety*, *176*, 258–269. https://doi.org/10.1016/j.ecoenv.2019.03.096.
Mishra, S., Yang, X., & Bahadur, H. (2020). Evidence for positive response of soil bacterial community structure and functions to biosynthesized silver nanoparticles: An approach to conquer nanotoxicity? *Journal of Environmental Management*, *253*, 109584. https://doi.org/10.1016/j.jenvman.2019.109584.
Mohd, M. A., Mark, E., Omar, S. A., & Alistair, B. A. (2018). Fate, uptake, and distribution of nanoencapsulated pesticides in soil–Earthworm systems and implications for environmental risk assessment. *Environmental Toxicology and Chemistry*, *37*(5), 1420–1429.
Nisha Raj, S., Anooj, E. S., Rajendran, K., & Vallinayagam, S. (2021). A comprehensive review on regulatory invention of nanopesticides in Agricultural nano formulation and food system. *Journal of Molecular Structure*, *1239*, 130517. https://doi.org/10.1016/j.molstruc.2021.130517.
Oliveira, H. C., Stolf-Moreira, R., Martinez, C. B. R., & Sousa, G. F. M. (2015). Nanocapsules containing atrazine toward maize plants. *Frontiers in Chemistry*, *3*, 61. https://doi.org/10.3389/fchem.2015.00061.

Pascoli, M., de Almeida, A. A. F., Almeida, J. A., & Fraceto, L. F. (2020). The potential of nanobiopesticide based on zein nanoparticles and neem oil for enhanced control of agricultural pests. *Journal of Pest Science*, *93*, 845–855. https://doi.org/10.1007/s10340-020-01194-x.

Peixoto, S., Henriques, I., & Loureiro, S. (2021). Long-term effects of $Cu(OH)_2$ nanopesticide exposure on soil microbial communities. *Environmental Pollution*, *269*, 116113. https://doi.org/10.1016/j.envpol.2020.116113.

Raliya, R., Saharan, V., Dimkpa, C., & Biswas, P. (2018). Nanofertilizer for precision and sustainable agriculture: Current state and future perspectives. *Journal of Agricultural and Food Chemistry*, *66*(26), 6487–6503. https://doi.org/10.1021/acs.jafc.7b02178.

Sahoo, J., Nanda, S., Mahapatra, C. R., Panda, D., & Kund, G. C. (2017). Lethal toxicity of deltamethrin and histological changes in the vital organs of fingerlings of *Labeo rohita*. *Journal of Fishery Sciences*, *5*(3), 506–513.

Santiago, E. F., Agathokleous, E., & Iavicoli, I. (2020). Understanding the interaction of nanopesticides with plants. In *Nanopesticides* (pp. 69–109). Springer International Publishing.

Sarlak, N., & Taherifar, A. (2017). Encapsulation of nanomaterials and production of nanofertilizers and nanopesticides: Insecticides for agri-food production and plant disease treatment. In *Nanoscience and Plant–Soil Systems* (pp. 481–498). Springer International Publishing.

Science, E., & John, S. (2012). Distinctive effects of nano-sized permethrin in the environment. *Environmental Science and Pollution Research*, *19*(2), 1107–1114. https://doi.org/10.1007/s11356-012-1161-0.

Sułowicz, S., Markowicz, A., Dulski, M., Nowak, A., Środek, D., & Borymski, S. (2023). Assessment of the ecotoxicological impact of captan@ZnO 35–45 nm and captan@SiO2 20–30 nm nanopesticide on non-target soil microorganisms–A 100-day case study. *Applied Soil Ecology*, *184*. https://doi.org/10.1016/j.apsoil.2022.104789.

Sun, Y., Zhao, X., Zhang, Y., Chen, S., & Jin, H. (2019). Nanopesticides: A great challenge for biodiversity? *Nano Today*, *29*, 4–7. https://doi.org/10.1016/j.nantod.2019.06.003.

Taghizadeh, R., & Mohammadkhani, N. (2017). Antifeedant activity of *Descurainia sophia* and *Thuja orientalis* extracts against *Tribolium castaneum* (Coleoptera: Tenebrionidae). *Journal of Crop Protection*, *6*(4), 487–495.

Van Gestel, C. A. M., Loureiro, S., & Zidar, P. (2018). Terrestrial isopods as model organisms in soil ecotoxicology: A review. *ZooKeys*, *162*, 127–162. https://doi.org/10.3897/zookeys.801.21970.

Villaverde, J. J., Sevilla-Morán, B., López-Goti, C., Alonso-Prados, J. L., & Sandín-España, P. (2018). Considerations of nano-QSAR/QSPR models for nanopesticide risk assessment within the European legislative framework. *The Science of the Total Environment*, *634*, 1530–1539. https://doi.org/10.1016/j.scitotenv.2018.04.033.

Vimala Devi, P. S., Duraimurugan, P., & Chandrika, K. S. V. P. (2019). Bacillus thuringiensis-based nanopesticides for crop protection. In *Nano-Biopesticides Today and Future Perspectives* (pp. 249–260). Elsevier.

Xu, C., et al. (2018). Emulsion-based synchronous pesticide encapsulation and surface modification of mesoporous silica nanoparticles with carboxymethyl chitosan for controlled azoxystrobin release. *Chemical Engineering Journal*, *348*, 244–254. https://doi.org/10.1016/j.cej.2018.05.008.

Yasur, J., & Rani, P. U. (2013). Environmental effects of nanosilver: Impact on castor seed germination, seedling growth, and plant physiology. *Environmental Science and Pollution Research*, *20*(12), 8636–8648. https://doi.org/10.1007/s11356-013-1798-3.
Zimmer, M., & Topp, W. (2002). The role of coprophagy in nutrient release from feces of phytophagous insects. *Ecological Entomology*, *27*(2), 1093–1099.

Index

Q

R